Performance Management

Performance Management

A business process benchmarking approach

Edited by

Asbjørn Rolstadås

Professor of Production and Quality Engineering,
University of Trondheim, The Norwegian Institute of Technology,
Trondheim, Norway.

SPRINGER-SCIENCE+BUSINESS MEDIA, B.V.

First edition 1995

Originally published by Chapman & Hall in 1995

ISBN 978-94-010-4531-5 ISBN 978-94-011-1212-3 (eBook)
DOI 10.1007/978-94-011-1212-3

A catalogue record for this book is available from the British Library

∞ Printed on permanent acid-free text paper, manufactured in accordance with ANSI/NISO Z39.48-1992 and ANSI/NISO Z39.48-1984 (Permanence of Paper).

CONTENTS

CONTRIBUTORS

Bjørn Andersen

Harald Bredrup

Reinholdt Bredrup

Ann-Charlott Pedersen

Karianne Prytz

Asbjørn Rolstadås

Tim Torvatn

All the contributors are affiliated to:
Norwegian Institute of Technology
University of Trondheim
Norway

PREFACE

This book serves as a textbook for an introductory course on performance management. It gives an overview over various aspects of managing performance of the modern enterprise by focusing on performance evaluation and measurement and performance improvement techniques. Most of the material is based on a thorough literature search and an extensive reference list has been included.

The book has been sponsored by the Norwegian productivity research program TOPP and by the COMETT program of the European Community Commission. It has been applied as the text for a continuing education course both within TOPP and the COMETT project APECE. It will also serve as part of a course material for a master's degree in technology management.

The book is aimed at an audience of business and technology oriented personnel at middle and higher management level in manufacturing industry. At the same time it is suitable as a textbook for business and engineering schools and colleges.

The book is organized in five parts discussing productivity and performance, performance planning, performance review, performance improvement and performance influencing factors. The authors have worked closely together to obtain a well coordinated text without overlap. They have provided a draft. This draft has been circulated for comments amongst the authors and amongst external experts. Based on their input the manuscript has been revised.

Eivald Røren and Einar Printz Moe, chairman of the board and program manager for the TOPP research program respectively, have also provided valuable input to the book.

I take this opportunity to thank each of the authors who have devoted much time and effort to accomplish this text. We all hope that the overview this book provides will turn out to be useful both to industry and academia.

Professor Asbjørn Rolstadås, Editor

Trondheim, January 15, 1994

PART ONE

Productivity and Competitiveness - an Introduction to Performance

1. THE PRODUCTIVITY TERM

Productivity - to some, the word prompts images of smoothly running processes and operations displaying efficiency to an extent that borders on sheer perfection and beauty. Others react with anger when someone mentions productivity. To them, productivity is equivalent with having to work harder for less pay, lay-offs and the like. What is this mysterious concept of productivity? How did it start and what has it come to mean?

1.1 FROM CRAFTSMANSHIP TO INDUSTRIAL PRODUCTIVITY

During the eighteenth century, the middle classes in Great Britain gained rapidly increased economic power, while a restructuring of agriculture created an excess of available labor. At the same time, Great Britain fought wars all over the globe that created and opened up a huge world market for finished goods. Together with the accelerated pace of technological development, these forces led to the industrial revolution in Britain at the end of the eighteenth century. Further technological advances and a number of inventions in turn caused a reorganization of production, from small groups to hundreds of people in large factories. This shift soon spread to the rest of Europe and the United Sates.

Although different industries and different countries developed with different speeds, attempts have been made to identify some characteristic stages in the development process toward modern, industrial production. Jaikumar (1988) denotes the first stage the English System of Manufacture, stretching from 1800 to 1850, where the ethos was accuracy. Parts were made one at a time, for one product, and to fit exactly together. Perfection was the objective, and the better the fit, the better the workmanship. Though productivity increased dramatically compared to earlier methods, manufacturing was still quite inefficient and labor intensive.

Around 1850, the American System of Manufacture evolved, based on both perfection and interchangeability of parts. Interchangeability relies on the existence of clearances between parts, where the greater the clearance, the easier it was to make parts interchangeable, as long as the parts stayed within defined limits that ensured functionality. This lead to the separation of operations from one another, such as systems for manufacturing large lots of parts and their inspection. Again, productivity leaped from the level of the last stage, but was still comparatively low.

Toward the end of the nineteenth century, Frederick W. Taylor realized that the machines of the time were more than capable, but worker-related activities were limiting the speed and efficiency that could be achieved by the machines. This realization lead to the stage denoted the Taylor System and scientific management, with focus on controlling human activities in order to increase labor productivity.

At this point, the transition from craftsmanship to industrial productivity was completed. Expensive machines were being installed that had to be kept running. To justify the high investments, machine utilization had to be high. As human activities limited the utilization rate, labor efficiency had to increase accordingly. Industrial productivity became synonymous with machine utilization and labor efficiency.

1.2 EARLY METHODS FOR INCREASING PRODUCTIVITY

Taylor and his ideas of scientific management

Each person's job should be broken down into elements, a scientific way to perform each element should be determined, and workers should be scientifically selected and trained to do the work in the designed manner. These are ideas and techniques almost exclusively focused on increasing the efficiency of the workers.

Frank and Lillian Gilbreth - time and motion studies

Through closely studying the workers performing their tasks, some basic motions were identified. Each operation was thought to consist of some sequence of some or all of these basic motions. The aim was to remove all waste motions and at the same time reduce the time it took to execute the necessary ones, all in order to increase efficiency. In some extreme cases, this quest for efficiency went too far. The output was produced in an extremely efficient manner, but was of very poor quality.

Henry Ford and the assembly line

The assembly line was a tool for further specialization of jobs and the removal of unnecessary motions, and thus further increased labor productivity.

The learning curve

The learning curve is not in itself a method for increasing productivity, but rather a principle that most of the early methods relied on for productivity gains. Figure 1.1 shows an example of a typical learning curve.

The idea is that a worker performing a specific task will as time passes learn to perform this task better or faster, and thus introduce productivity gains into the process. As the curve illustrates, the learning effect is slow in the beginning. Later on, the slope of the curve increases rapidly until there is little left to gain and the curve flattens.

The Hawthorne studies

The researchers at the Hawthorne plant felt that if the best work environment could be determined (the same way the best way to perform the job could be determined by scientific management), the workers would be more efficient and become less tired. The studies revealed that viewing all workers as rational and economic beings was inadequate, and a social view of workers evolved. Managers started addressing their employees' feelings and attitudes and potential effects on productivity.

It is therefore evident that the sole purpose of the traditional productivity-enhancing principles was to increase the amount of work performed per hour or rather per £ or \$. Are these principles suited for today's world of industrial manufacturing? We do not think so. During the next chapters, we will try to explain why.

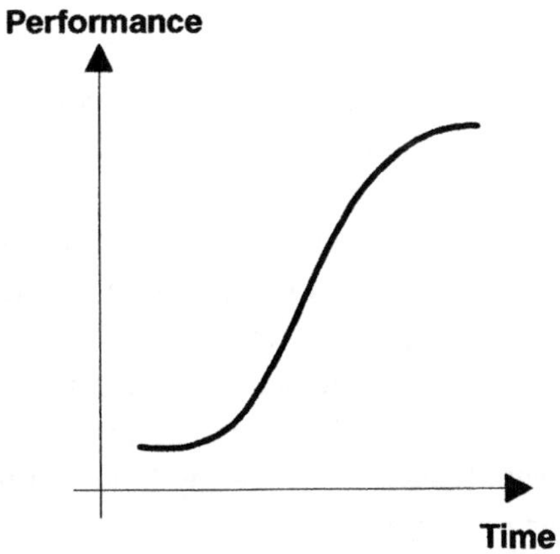

Figure 1.1 The learning curve, also called the S-curve.

1.3 THE PREVAILING PRODUCTIVITY PARADIGM

As this brief historical overview has shown, the prevailing rule has been the unusual rather than ordinary conditions for promoting world trade since the productivity term came into use. During the nineteenth century, people were for the first time in history enabled to buy finished goods, fresh out of modern factories. The slope of technological development was steeper than it had ever been before, leading to whole new products more and more frequently. People bought these products, and management were challenged to increase worker productivity, replace obsolete work methods and equipment with newer, more expensive equipment while developing and introducing new products. All in all, the key focus was efficiency: efficiency of workers and efficiency of machines.

The next century increased even further the slope of the technological development curve, but also brought on two world wars. For decades after their ending, these wars created truly extraordinary conditions for industry. The world markets were almost insatiable and would consume products of almost any quality. Again, the challenges were to drive the technological development ahead, mainly on the premises of the developers. Very rarely were the customers asked what they wanted. Still, companies survived this negligence of customer needs, mainly because consumers were more oriented toward quantity than quality. Based on the belief that success with innovative products could offset poor performance in traditional operations, productivity was still synonymous with efficiency, i.e., labor productivity and machine utilization.

It is evident though that the circumstances facing companies up to only a few decades ago were highly extraordinary. In fact, such circumstances dominated for so long that many came to view them as ordinary, and thus, the sciences of management were developed according to them. That is one of the reasons why this particular productivity paradigm has proved to be so suitable for survival.

Another reason for the prevalence of the efficiency paradigm is the immense support it has enjoyed from accounting systems and bonus and incentive systems. The principles for financial and accounting systems were developed during those extraordinary circumstances we discussed just recently. These principles are still very much alive today and in use to an extent one would hardly believe, considering how fast the systems of other areas have been made obsolete.

Key words have become 'efficiency' through cost cutting and 'outsourcing' to the extent that they very often undermine revenue-generating activities and core competencies of the company. Furthermore, focus is on figures measuring some sort of output per employee, regardless of whether the output is of satisfactory quality and actually wanted by the customer. Emphasis is on machine utilization that has kept machines going, generating waste most of the time, as the parts were not needed at the time and often obsolete before they were needed. Meeting budgets that often meant reduced investments, both in equipment and training of personnel, as well as meeting quotas that often lead to shipping parts of poor quality at the end of accounting periods. Bonus and incentive systems have had some of the same effects on worker output level, the key goal being absolute amount of output, not amount of output satisfying quality demands.

All of this has nurtured the efficiency paradigm, which in turn has led to a high degree of specialization within companies. This has caused high barriers between departments making communications difficult and slowing or even stopping the voice of the customer from passing through the company. Very often, the key basis for increased efficiency has been movement down the learning curve, without any concern for continuously improving ones' methods of operation. When the slope of the learning curve has started decreasing, further efficiency gains have been realized depending on technological breakthroughs. The result has been a very bad climate for listening to the customer and the need to continuously improve to keep up with his demands.

Another traditional and very strongly held belief that falls under the efficiency paradigm is the perceived trade-off between price, or cost, time and quality. As quality was believed to cost more and take longer to manufacture, it has often been sacrificed for the sake of reduced cost and efficiency.

It might seem we are painting an extremely ugly picture here, but it is a fact that the prevailing productivity paradigm has caused quite a handful of misunderstandings and wrongly based perceptions that are only now starting to change. By putting words to some of these processes, we hope to speed up the transition to the new productivity paradigm.

1.4 TRADITIONAL PRINCIPLES AND METHODS

Different definitions of productivity

Traditionally, productivity has been defined in terms of the relationships between the inputs and outputs of a productive system, most commonly expressed as a ratio. The most general definition is thus:

$$\text{Productivity} = \frac{\text{Output}}{\text{Input}}$$

This basic expression can in turn be adapted to the appropriate level of aggregation, that is, unit of analysis, and to the actual factors of output and input that are being measured. Finding the right expression when measuring productivity at worker or group level is often a task that is best done by those involved and exclusively fitted to the individual situation. Although some general definitions have been suggested for industries as a whole or for particular types of companies, there exist very few ready-made lists to pick from.

For measuring productivity at country or region level, however, several different definitions are in frequent use. OECD publishes measures for productivity, or trends in performance, on a national level through measuring trends in apparent labor productivity. This is defined as Production/Employment. As most countries track indeces for almost everything, an easy obtainable measure is Production Index/Employment Index. The Norwegian Central Bureau of Statistics (NCBS) employs two definitions:

1: Gross Production Value / (Input Goods (in value) + Wages)

2: Gross Production Value / [Input Goods (transformed into man-years) + Labor (in man-years)]

These definitions all reflect the fact that productivity has traditionally been a subject that economists have been concerned with, and at a country level. The main outcome has been country comparisons for use in different international forums.

Another point to be made is that when deciding upon which definition to use for a particular situation, it is important to bear in mind that different definitions can cause extremely varying results. Figure 1.2 shows how a few common expressions give totally different results. All of the curves express trends in labor productivity in Norwegian Technology Industries, using 1980 as reference year.

Definitions of productivity are to a large extent linked to corresponding models for measuring productivity. Principles and models for measurement will be dealt with at a later stage, so the rest of this section will, without

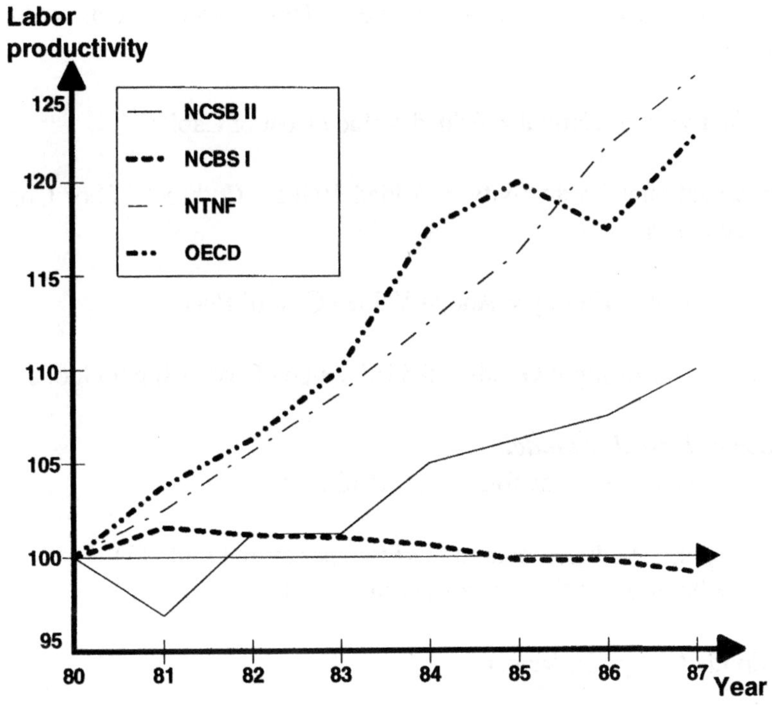

Figure 1.2 The development of labor productivity in Norwegian technology industries according to different definitions of productivity.

going into any detail, give a brief overview of some measurement models and connected definitions.

Total and partial productivity

When dealing with productivity on a micro level, i.e., inside companies, it is common to separate between the total productivity of the company as a whole and the productivity of different input factors. One definition of total productivity is:

> Total Productivity = Added Value / [(Direct Labor Costs + Social Costs) + (Indirect Labor Costs + Social Costs) + Cost of Capital]

Splitting the total productivity into partial productivity measures, one is better able to determine for which input factors the utilization applies satisfactorily and for which it needs to be improved. Some areas for which the partial productivity can be measured include:

Labor Productivity = Added Value / (Direct Labor Costs + Social Costs)

Productivity of Capital = Added Value / Cost of Capital

Organizational Productivity = Added Value / (Indirect Labor Costs + Social Costs)

Productivity of Energy = Added Value / Cost of Energy

Productivity of Input Goods = Added Value / Cost of Input Goods

Quantity-price-value model
This model is based on the following definitions:

Revenues = Produced quantity x Price per output unit (1)
Cost = Input quantities x Price per input unit (2)

If we divide (1) by (2), we get:

Revenues / Cost = (Produced quantity / Input quantities) x (Price per output unit / Price per input unit)

or

V1 / V2 = (Q1 / Q2) x (P1 / P2)

or

Change in profits = Change in productivity x Change in price/cost recovery

as shown in Figure 1.3. There are numerous software packages available for measuring and presenting data from this model.

1.5 THE NEW PRODUCTIVITY PARADIGM

As the rebuilding after World War II approached completion and the demand caused by years of non-normal production was met, ordinary,

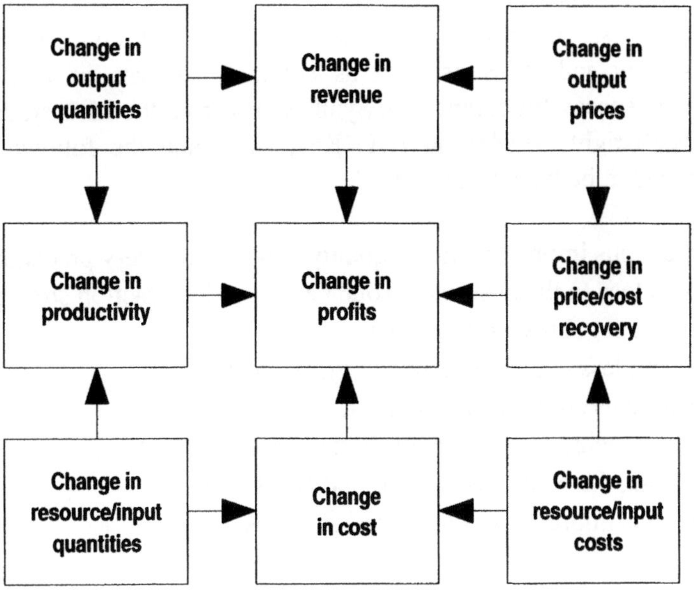

Figure 1.3 The principles of the quantity-price-value model

competitive conditions were introduced for the first time in a long, long time. The customers were no longer dazzled by any new product that was introduced, and the market changed from being the seller's to becoming the buyer's. The customers could compare different offers and choose among them. Buying criteria other than price came into consideration, criteria that the traditional productivity paradigm was no longer able to satisfy. Simply to efficiently manufacture products without really caring for the customer is no longer enough.

The modern customer, both industrial and consumer, wants and expects the right products, of leading edge technology, with outstanding quality, delivered with minimum lead times and at a minimum price. These are the new demands that the new productivity paradigm will have to support. The philosophies of Just-in-Time, Total Quality Management and others have shown that the traditionally perceived trade-off between price, quality and time is no longer valid. There is in fact an unconditionally positive connection between the three. These philosophies have also highlighted new areas of opportunities and emphasized the need to think holistically and for long-term benefit.

Several international studies, mainly initiated by the decreasing US competitiveness, have both identified key areas of concern in failing companies or industries as well as common patterns of thinking and behavior in successful companies. Synthesizing from the MIT study, Hayes and Wheelwright (1984) and A.T. Kearney, found the following issues have proved to be keys to success:

- simultaneous improvements in quality, cost and delivery precision;
- staying close to the customer, Total Customer Satisfaction strategy;
- closer relationships with suppliers;
- flatter and less departmentalized organizations;
- innovative use of human resources;
- the use of technology to achieve strategic advantages;
- developing the employees' skills and capabilities;
- continuous improvement and strategic leaps.

All of this proves the need for trashing the old productivity paradigm of efficiency alone and moving on to a much wider productivity paradigm.

1.6 FROM EFFICIENCY TO EFFECTIVENESS: MEASURING PERFORMANCE

The key to the new productivity paradigm is to accept that productivity, in the sense of efficiency, is only one of several important dimensions of the organization that must be measured and improved. We need to move on to include in the equation both effectiveness as well as some of the softer and more intangible dimensions. The new productivity paradigm is actually the performance paradigm, where we actually measure the entire performance, including productivity, of the organization. We will in the following pages present some of the modern and, in our opinion, much more effective definitions of 'productivity', i.e., performance.

Lawlor (1985), although still using the term productivity, provides one of the first wider definitions. He identifies five aims any organization should strive for, and defines productivity as a measure of how efficiently and effectively the organization satisfies these dimensions. The aims are as follows:

1. Objectives, or to what degree the organization's main goals are achieved.
2. Efficiency, the traditional measure of productivity, how efficiently resources are used to generate outputs. However, Lawlor includes the condition that the generated output is actually needed by the market.
3. Effectiveness, where Lawlor means a measure of potential, what is achieved compared to what is possible.
4. Comparability, the productivity score of the organization in question compared to other organizations or industry/country averages.
5. Trends, the performance record over time, whether it remains static, increases or decreases.

Sink (1985) and Sink and Tuttle (1989) suggest that the performance of an organizational system is a complex interrelationship between the following seven performance criteria:

1. Effectiveness, doing the right things, at the right time, with the right quality, etc.
2. Efficiency, doing things right.
3. Quality, where quality is an extremely wide concept.
4. Productivity, in the traditional sense.
5. Quality of work life.
6. Innovation.
7. Profitability/budgetability, the ultimate goal for any organization.

The last example of a modern productivity definition we will include, is one suggested by A.T. Kearney in Byrne and Markham (1991). Their definition is highly customer focused and cuts through the use of criteria to simply state productivity as:

> *Using the combined resources of all the participants in the supply chain in the most efficient way to provide high quality, cost-effective customer service.*

This means that first, companies must ensure they provide the customer with what he wants (doing the right things). Then, they should seek ways to improve the efficiency of doing so (doing things right).

Putting together these definitions, they cover some quite comprehensive grounds, but still we feel there are one or two dimensions of the organizational performance that has not, at least not clearly enough, been emphasized. Agreeing with Imai (1986) that there is a basic difference between an innovation strategy and a continuous improvement, or KAIZEN strategy, and it is the latter element which is often missing. What is really needed is a combination of these two concepts. Continually improving is the basis for reducing waste, both time and money, improving quality, and assuring the ability to deliver the best possible offer at lowest total price for the customer. Innovation can dramatically enhance the continuous improvement approach by occasionally providing breakthrough improvements and facilitate necessary changes to respond to changing surroundings. We therefore believe that the criteria of continuous improvement should be added to the presented definitions, as shown in Figure 1.4. The figure depicts a generalized value-adding system, consisting of an upstream system, i.e., a supplier, one's own value-adding activities, and output from the process to a downstream system, i.e., a customer. Employing the principle of continuous improvement means to enhance the performance of the entire value-adding system.

Furthermore, as it is becoming increasingly more evident what damage we are inflicting on this planet, there are compelling reasons for including environmental impact in the performance set. Not only should we do this because customers are growing aware of this fact, and that this matter is able to stir some quite forceful emotions, resulting in boycotts, but also for our own sakes. It is crucial if we still want to inhabit the earth. It is possible that this criterion should be viewed as part of the quality criterion, as it has a bearing on the overall quality of the offering, defining quality in the broadest sense. One could also argue that whether the offer has a positive environmental profile is part of the organization's effectiveness, i.e., whether it is accomplishing, putting out, the right things. Regardless of whether environmental impact is made an individual criterion or included in effectiveness or quality, we will strongly emphasize the growing

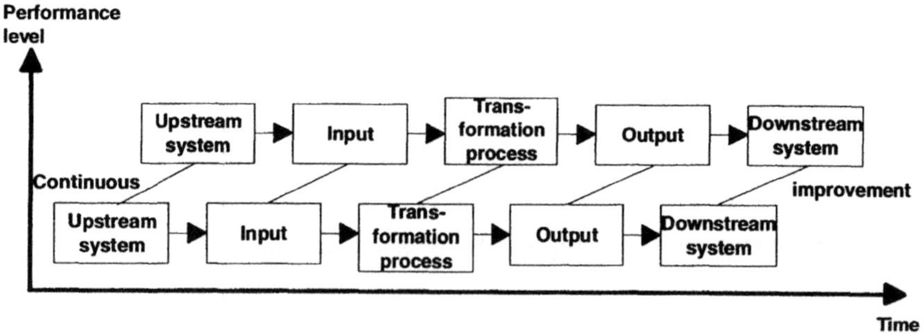

Figure 1.4 Illustration of the performance dimension continuous improvement.

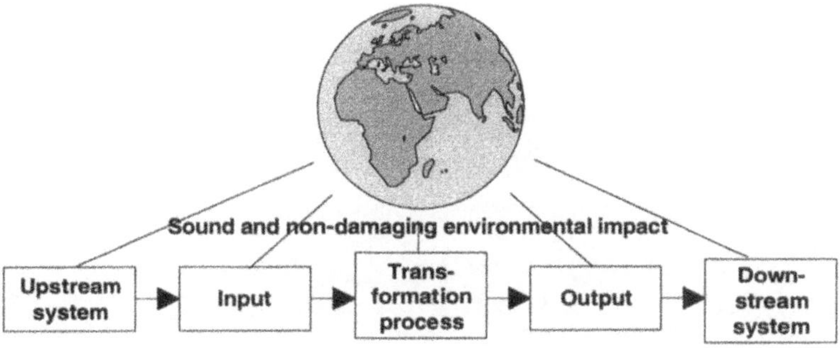

Figure 1.5 Illustration of the performance dimension environmental impact.

importance of including it in your performance. As it is shown in Figure 1.5, simply having environmentally sound processes within your own domain is not enough, you have to ensure that every link in the supply chain is equally good. Carrying one weak link will eventually be discovered by the customers and bring discredit to the end product.

We are confident that there exist numerous other definitions of productivity/performance that might express the message we want to get through just as well as these do. However, we feel that these pretty much cover our case. They are quite clear in language, and range from comprehensive criteria to a simple and direct statement. It is up the reader to interpret performance the way that makes most sense to you. We just hope we have managed to make you see the need for the new productivity paradigm.

2. PRODUCTIVITY STUDIES

Competitiveness is heavily stressed both on a local and a national level in all countries. Competition becomes more and more global through international agreements like GATT. Competitiveness is a necessity in order to survive in the new age of competition. Productivity is closely related to competitiveness as a measure for development. No single metric for competitiveness exists, because this is a multi-dimensional quantity.

2.1 INTRODUCTION TO THE WORK OF PORTER

This chapter is a brief introduction to the work of Porter. His book '*Competitive Strategy: Techniques for Analyzing Industries and Competitors*' (1980) introduced three strategies to create competitive advantage: cost leadership, differentiation and focusing. His next book '*Competitive Advantage: Creating and Sustaining Superior Performance*' (1985) introduced the value chain as a means to diagnose competitiveness of companies. The value chain of a company is a part of the total flow of activities described in the value system. Porter's latest work '*Competitive Advantage of Nations*' (1990) is an international study of competitiveness. He studied why companies in some countries are competitive in special industries and how global strategies influence this ability.

2.1.1 Competitive strategy

Porter states that strategies at the top level to create competitive advantage could be divided in three categories. A company has to choose one of these strategies. By trying to combine different strategies they risk falling between two schools and eventually loose competitiveness. According to Porter the three generic strategies are:

- Cost leadership
- Differentiation
- Focusing

We will not go into detail about how to develop strategies, but give an overview of the most important terms within the scope of this book.

Cost leadership

Cost leadership means that the company aims at achieving the lowest production costs within the industry. Sources for competitive advantage could be:

- Scale advantages
- Protected technology
- Advantageous access to resources (energy, goods, etc)

A typical cost leader has a wide range of products and could operate in related industries. To become a cost leader identification and exploitation of all possible opportunities to gain cost advantage is necessary. Relying on learning curve development is not enough to stay competitive. Typically a cost leader offers plain standardized products to keep costs to a minimum.

Cost advantage could be realized as a higher profit than competitors provided that the market perceives the products comparable to products from the competitors. If the competitors have been able to differentiate their products, the cost leader has to reduce the price to compensate for the disadvantage.

Differentiation

Differentiation means that the company tries to be perceived as unique by the market regarding criteria highly valued by the customers. Differentiation criteria will differ from industry to industry and could be based on different aspects like:

- Product (quality level, functionality, user-friendliness, etc)
- Distribution (distribution channels, delivery times, etc)
- Packaging (exclusive packaging, extra service, etc)
- Marketing (image, guarantee, service, etc)

A differentiation criterion has to be perceived as extra value added by the customer to be real. Competitive advantage is achieved only if perceived extra value added exceeds the cost of providing the differentiation. There is a close connection between cost and differentiation and the relative cost position could not be ignored. A differentiation advantage could be eaten up by poor cost competitiveness in other areas.

Focusing

Focusing means that the company should choose a narrow niche within the industry. This is obtained by choosing a segment or group of segments within the industry. The strategy is to serve these segments so good that other competitors are excluded. Competitive advantage is achieved by focused attention to the niche. Two strategies are possible:

- Cost focusing
- Differentiation focusing

A company with a high market share in a niche could achieve scale advantages and offer a lower price. A focused company could easier identify the special needs for the niche and provide differentiation that is impossible for companies serving the whole industry.

2.1.2 Levels of competitive advantage

Porter has studied competitiveness on different levels. He has developed tools and methods to analyze competitiveness in a value chain, a value system, clusters and nations. We will discuss the most common terms introduced in this work.

National

Porter (1990) has identified four broad attributes of importance for competitiveness of a nation. These four attributes are tied up in a network with relations in all directions like a diamond shown in Figure 2.1. The 'diamond' is an analysis tool in the strategic planning process.

Factor conditions. Access to factors of importance for production. Standard theory says that countries with factor advantages will export in these industries. It is important to note that these factors are created, not inherited, which points out creation and upgrading of factor conditions to be a critical focus for every government. Five different types of factor conditions are identified:

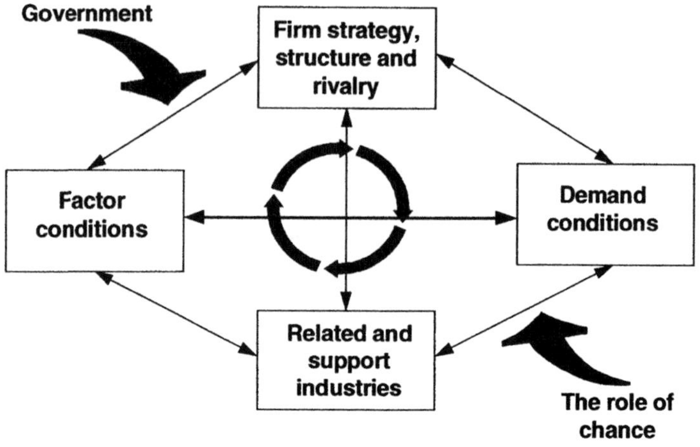

Figure 2.1 The determinants of national advantage

- Human resources
- Physical resources
- Knowledge resources
- Capital resource
- Infrastructure

Factor conditions are divided in basic and advanced factors. Basic factors are often inherited whereas advanced factors have to be developed. Advanced factors are more important than basic and more difficult to develop and therefore harder to copy. Porter divides between generalized and specialized factors. This division is due to dynamic development. Today's specialized factors are tomorrow's generalized factors.

Demand conditions. Characteristics of home market demand for the industry's product or service. However, home market does not automatically mean domestic market. The EU is a natural home market for many European companies.

- Home market composition. The mix and character of home buyer needs are the most influencing factor of home demand. Porter makes a tri-partition of home demand composition. The first, segment structure of demand, reflects the fact that some segments are more global than others. Sophisticated and demanding buyers are a valuable asset for a company. If buyers in the home market are among the world's most sophisticated and demanding buyers for the product, the company probably has better access to the most advanced buyer needs than their competitors. Less sophisticated buyers are followers and will adapt new needs later. Anticipatory buyer needs mean that changes in home demand could be an early warning of buyer needs that will eventually be adopted in other markets.

- Demand size and pattern of growth. Influencing factors are said to be the size of the home demand, number of independent buyers, rate of growth of home demand, early home demand and early saturation.
- Internationalization of domestic demand. Factors like mobile or multinational local buyers and influences on foreign needs are mentioned by Porter.

Presence of sophisticated and demanding customers are very important for development of competitiveness.

Related and supporting industries. Presence of internationally competitive home market based supplier industries or related industries. Access to a local competitive supplier industry and related industry is an important determinant for competitiveness:

- Competitive advantage in supplier industries benefit other companies in several ways:
  - close, quick and preferred access to cost effective input
  - close coordination into the value chain with first class deliveries
  - a continuous innovation and upgrading process, early access to new products/processes and information and possibility to give feedback is perhaps to most valuable benefit.
- Competitive advantage in related industries. Competitive related industries could initiate development of new competitive industries. This gives an opportunity for information exchange, transfer of technology and more rapid identification of new opportunities.

Firm strategy, structure and rivalry. How companies are created, organized and managed as well as the nature of local rivalry. Management styles differ from country to country and a special style could be successful in specific industries. Differences between countries could be seen due to differences in for instance education, attitudes, culture, religion, language skills and political aspects. Porter points out three major aspects:

- Strategy and structure of domestic companies. No country has a uniform way of managing companies, but by studying different countries Porter has identified similarities. A nation's industry structure influences the choice of strategies. Focusing strategies are for instance most widespread in countries with mainly small- or medium-sized enterprises.
- Goals. Porter has identified sharp differences in the goals that companies are using to describe performance and are using to motivate employees and managers. Nations will succeed in industries where these goals and motivations are aligned with the sources of competitive advantage.
- Domestic rivalry. Porter has strong evidence of the correlation between vigorous rivalry and the creation and persistence of competitive advantage. Rivalry means pressure to improve and innovate. Rivalry, demanding customers, and so forth, helps create a change-oriented organizational culture necessary for long-term survival.

The 'national diamond' illustrated by Porter in Figure 2.1 contains two other aspects that could influence all four determinants of national competitiveness. The role of chance is independent of nations, but national attributes influence the national exploitation of events. Events can create discontinuities that create possibilities to change the diamond. Government is not the fifth determinant, but influence the four others.

Porter describes the diamond as a system. Success in international competition is a result of advantages in the diamond. This is not static, but a dynamic model of the determinants of competitiveness of a nation. National advantage is continuously changing and the diamond is only a descriptive model. Porter describes the diamond as a spiral. All determinants have mutual influence on each other and could reinforce the advantages created by one factor. The descriptive power of the model decreases when the developments become more rapid. Porter has found and described important relationships between the different determinants of the diamond. We will not go into detail. However, it could be mentioned that Porter points out elements of special importance; domestic rivalry and geographic concentration. Domestic rivalry promotes competition and continuous improvement that helps upgrading the entire national diamond. Geographic concentration stimulates the interaction within the diamond.

Cluster

A cluster is a subgroup of the nation and consists of industries related by different links. Competitiveness of a nation is not likely to be equal for all industries. Porter found that a nation's successful industries were usually linked through vertical (buyer/supplier) or horizontal (common customer, technology, channels, etc) relationships. He also found that successful clusters were geographically concentrated. Clustering is favored by all the positive influencing mechanism within the diamond. A cluster becomes mutually supporting and a means to overcome introspective satisfaction and conservatism. Development opportunities are best with perfect exchange and flow of information regarding customer needs and technology within the cluster. This is however often limited by conflicting interests within the cluster, but could be stimulated through:

- Stimulating factors for information flow:
  - personal relationship (from school or organizations etc.)
  - relations through academic milieus/organizations
  - relations because of geographic nearness
  - trade organizations representing the cluster
  - behavior pattern for long-term relationship.
- Sources for shared goals and compatibility within the cluster:
  - family or quasi-family relations between the companies
  - common ownership
  - common managers/directors of board
  - national patriotism.

Value chain

Competitive advantage could not be fully understood by just looking at the company as a whole. Competitive advantage is created by the different activities. Each activity consumes resources and influences the cost efficiency of the company. Performance of each activity could be a differentiation criterion. However, some activities are more important than others. A company cannot be world class performer in every activity.

The value chain shows the total value and shows value activities and margins. Value activities are the building blocks a company uses to create a product with value for the customer. Margin is the difference between the total value and the sum of costs for performing all activities. Porter divides the activities in two main groups as shown in Figure 2.2.

- Primary activities - activities necessary to manufacture, sales, transfer of the product to the customer and after sales assistance.

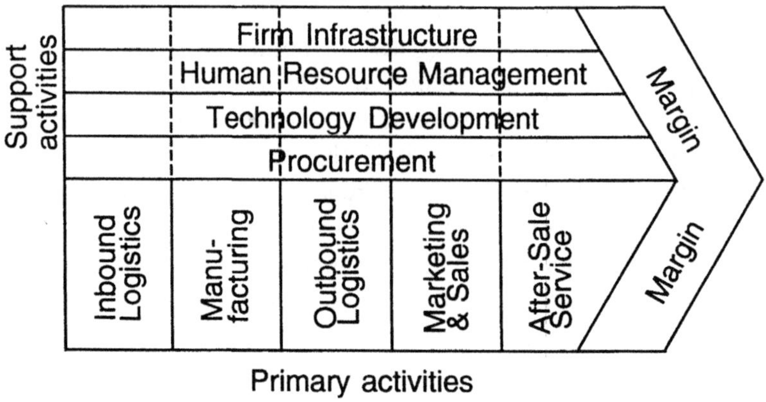

Figure 2.2 Value chain

- Support activities - activities that support the primary activities and each other through procurement of the resources, technology, human resources and different functions all over the company.

Porter has identified five standard types of primary activity:

- Inbound logistics includes activities like receiving, storing and distributing factor inputs for the product.
- Manufacturing includes physical transformation activities of factor inputs into products.
- Outbound logistics includes activities like collecting, storing and physical distribution of finished products to customers
- Marketing and sales include activities for making products available for customers and persuading them to buy them.
- After sales service includes activities to improve and maintain the value of the product.

The standard set of support activities contains four categories:

- Procurement which is the function to provide factor inputs that are used in the value chain of the company.
- Technology development which includes efforts to improve both products and processes.
- Management of human resources which includes activities like recruiting, employing, educating, developing and rewarding all kinds of employees.

- A company's infrastructure which includes activities like administration, planning, finance, accounting, legal matters, communication with authorities and quality management.

However, each category of primary and support activities contain three subgroups of activities:

- Direct activities which are directly involved in adding value for the customer like manufacturing, assembly, marketing, etc.
- Indirect activities which are activities that make it possible to run the direct activities continuously and includes maintenance, planning, administration, etc.
- Quality assurance which is to ensure quality in all activities through control, inspection, testing, revision, checking, correction and rework. Quality assurance is not the same as quality management.

Every company has its own unique value chain and this standard set is only a guideline. Like other modeling methods, identification and definition of a company specific model is a difficult but very important task.

Value system

A value system is a chain of several value chains as shown in Figure 2.3. This example illustrates several roles in a total value system. The actors within the system could be part of several value systems. The nature of the linkages between the different value chains depends on their role in the total value system. Final product assembler or original equipment manufacturer (OEM) has a key role in the total value system. An OEM is most probably the 'owner' of the product and customer's voice has to be recorded here. OEM's then have a responsibility to educate and manage the whole value system, because they are the only ones with the power to change attitude and performance of all value chains. An old proverb says that no chain is stronger than the weakest link. This is also true for a value system. If flexibility is regarded important by customers', the OEM has to ensure that the required performance is understood and met by his first-tier supplier. They also have to ensure that second-tier supplier offers flexibility with their performance.

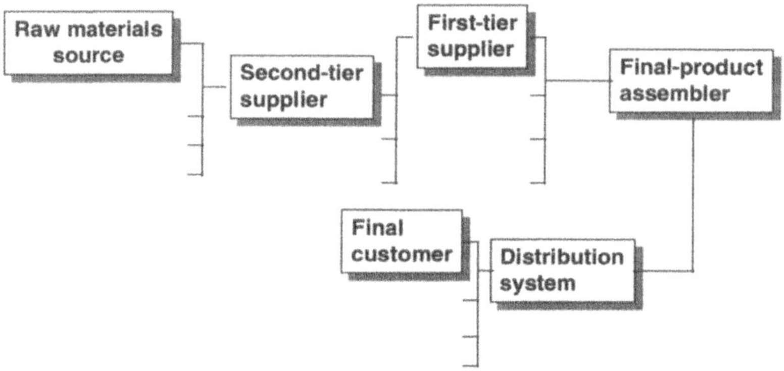

Figure 2.3 An example of a value system

2.2 MIT STUDY 'MADE IN AMERICA'

The growth in productivity in America has decreased during the past 15 years. It is now considerably lower than it has been for at least two decades and is fallen behind the productivity improvement rate in several Western European and Asian nations. Many proud American industries, including automobiles and steel, have lost much of their market share both at home and abroad. In a few industries such as consumer electronics, it is nearly impossible to find any American brand on the market.

With this reality in mind, Massachusetts Institute of Technology (MIT) established in 1986 the MIT Commission on Industrial Productivity to study the nation's industrial performance. The Commission sought first to weigh the evidence supporting the optimistic and the pessimistic views. This was to find out whether there were any pervasive weaknesses in US industrial practice and their causes. They also looked at changes in the international economic system to consider how those changes might alter the requirements for successful performance in the future. The ultimate aim of the study was to formulate a set of recommendations that would help the US industry to sustain strong growth in productivity. The study is named '*Made in America*' and has later been succeeded by many studies in other countries to prove the findings of MIT. It covered eight different industries:

- automobiles
- chemicals
- commercial aircraft
- consumer electronics

- machine tools
- computer and office equipment
- steel
- textiles.

Different teams were formed from members of the MIT community to evaluate its industry with respect to efficiency, product quality, innovativeness, adaptability and other aspects of performance. The teams compared American practices with those abroad and contrasted successful firms with failures. More than 200 companies in the US, Europe, and Japan were visited and almost 550 knowledgeable practitioners and analysts in industry, government, organized labor, and universities were interviewed.

In all industries studied, the marked share of imports had increased while the exports had decreased over the previous 15 years. The results showed that the US had lost in productivity development to Japan in all the industry segments studied except chemicals and commercial aerospace.

Even so, the study was optimistic in its conclusion and stressed the need for focus on manufacturing technology. The later development has shown good productivity improvement of the US industry, especially in the automotive industry. The study has led to a changed view on the importance of manufacturing technology and many US universities are now offering a manufacturing technology program.

2.2.1 Weaknesses in productivity performance

Comparing the observations from the different industries, recurring patterns of weaknesses in productivity performance were found. The commissioners discovered six interrelated patterns of behaviour that can be said to be the main reasons for the poorer productivity development in US. These were:

- obsolete strategies;
- too short an horizon for investments;
- technological weaknesses in development and production;
- inappropriate use of human resources;
- insufficient cooperation in the organization;
- conflict of interest between authorities and industry.

Outdated strategies

In several industries the study showed that managers and workers were so attached to the old way of doing things that they could not understand the new economic environment. The study reveals two main elements of past practice that are still common in US industry. These are mass production of standard commodity goods and parochialism.

Mass production was the driving force behind American postwar productivity development, but is now not an appropriate model to succeed in today's competition. While Detroit's car industry still sticks to the mass production system, the success of the Japanese car industry has succeeded in providing different products for each segment of the market. They have developed technologies that allow them to reduce the volume of production and improved the time to market of new products. The Japanese have emphasized highly skilled workforce as well as flexible automation. They have moved closer to both the customer and suppliers and have improved the quality and reduced the costs.

For decades American industry found a seemingly unlimited and uncontested outlet for their products in their own domestic market. The home market was large, unified, and familiar, while the foreign markets were not attractive. Foreign markets and competitors were largely ignored. Today American industry still suffers from their parochialism. The global business strategies of European and Asian firms have fundamentally changed the competition situation, both in the US market and abroad. Parochialism has also blinded Americans to the growing strength of scientific and technological innovation abroad and the possibility to adopt the advantages in US industry.

Short-time horizons

The Americans are more likely to concern themselves about short-term outcomes of their investments than their rivals in other nations. One factor is that US firms must often pay more for the use of capital and thereby must seek quicker payoffs. Nevertheless, the Commission suggest that these problems are also due to the practices of financial institutions and corporate managers and sometimes the risk perceived to be associated with the policies of the US and other governments.

Technological weaknesses in development and production

Even if the US is still the leader in basic research, the companies increasingly find themselves lagging behind their foreign rivals in the commercial exploitation of inventions and discoveries. The major enabling technological advances on products such as radios, color televisions and videocassette recorders were first made in US, but are now dominated by foreign manufactures. Earlier in the century, the US industry dominated the early stage of the product cycle in most industries and could move to the next generation of products before the competitors were able to acquire the technology to keep up with the US firms. Today the industrial nations of Europe and Asia have greatly expanded their technological capabilities and can rapidly capitalize on promising technological discoveries made anywhere in the world.

In American industry, several shortcomings have become evident in the performance in developing new products, engineering, and manufacturing. The industry studies pointed out the following factors:

- American companies find it difficult to design simple, reliable, mass-producible products.
- They often fail to pay enough attention at design stage to the likely quality of the manufactured product.
- Their product development times are excessively long.
- They pay insufficient attention to manufacturing processes.
- They take a reactive rather than a preventive approach to problem solving.
- They tend to underexploit the potential of continuous improvement in products and processes.

Neglect of human resources

The research on productivity and the quality of work force suggest that the origins of the problem with low labor productivity do not lie in the disappearance or weakening of basic American values and capabilities, but in the educational institutions. To keep up with their foreign rivals, the education system must be improved. The study blames the formal schooling for not providing an adequate education basis, but also points out training in the firms as important.

Failures of cooperation

Underdeveloped cooperative relationships between individuals and between organizations stand out in the industry studies as obstacles to technological innovation and the improvement of industrial performance. The lack of cooperation appears at several levels. This includes cooperation between individuals and groups within firms, between firms and their suppliers or their customers, among firms in the same industry segment, and between firms and government. The relations between management and labor are also not satisfactory.

Government and industry at cross-purposes
Firms operate in an environment shaped by federal macroeconomic policy and by a variety of other government policies concerning such matters as education and training, research and development, national security, economic and social regulation, and the nation's economic infrastructure. The industrial performance depends heavily on incentives and constraints created by these various governmental actions outside the macroeconomical sphere.

2.2.2 Patterns of best industrial practice

The Commission also identified firms that are responding successfully to the opportunities and constraints of the new environment. Several similarities among the best practice firms were found. These include:

- Focus on simultaneous improvement in cost, quality and delivery;
- Closer links to customers;
- Closer relationship with suppliers;
- Effective use of technology for strategic advantage;
- Less hierarchical and less compartmentalized organizations for greater flexibility;
- Human resource policies that promote continuous learning, teamwork, participation and flexibility.

Simultaneous improvement in quality, cost and delivery
It is not surprising that high quality, low unit cost, speed in moving new products from the design stage to market and speedy and reliable delivery of products and services are all fundamental virtues in any business organization. However, they found, that the best-practice firms put particular emphasis on simultaneous improvements in quality, cost and speed. The best companies have made significant improvements in all three properties.

Staying close to the customer

All of the successful firms in the study are making an effort to develop closer ties to their customers. They pick up more differentiated signals from the market and respond to different segments of demand. They also increase the likelihood of rapid response to shifts in the market. Successful niche manufactures are staying in close and continuous contact with their customers so that they rapidly redeploy their productive resources to new users. Similarly, the most successful high volume manufactures combine a continuing emphasis on economies of scale with great flexibility, shorter production runs, faster product introductions and greater sensitivity to the diverse needs of their customers.

Closer relations with suppliers

The closer and more tightly co-ordinated relationships with suppliers appear in all of the best-practice firms in the study. Coordination with the external firms is crucial in cutting inventories, in speeding up the flow of products and in reducing defects.

Using technology for strategic advantage

The industry studies revealed a trait common to the firms that are most successful in transferring scientific discoveries and new technologies into production and the marketplace. They have integrated technology choices into the rest of their business planning, including strategies for manufacturing, marketing and human resources.

Flatter and less compartmentalized organizations

In all successful firms in the study, the trend is toward greater functional integration and fewer layers of hierarchy, both of which promote greater speed in product development and greater responsiveness to changing markets. A flatter hierarchy generally enhances organization flexibility and broadens the responsibility associated with a particular job. It also promotes closer relationships with customers. The customers can often speak directly to the responsible production group instead of going through the sales department.

Implementation of cross-functional teams and simultaneous (concurrent) engineering has given good results. The successful companies also point out that it is essential to have a strong project manager who is capable of unifying the functions of the various team members and that the groups have skills in problem solving, teamwork and negotiation.

Innovative human resource policies

Best-practice firms have recognized that improvements in quality and flexibility require levels of commitment, responsibility and knowledge on the part of the work force that cannot be obtained by compulsion or cosmetic improvements in human resource policies. The workers must understand the overall production process and their wages, job security or profit-sharing arrangements must give them a sense that they have a stake in the firm's future. Quality is the output of an entire production system, and not the result of an organizational gimmick.

2.2.3 Imperatives for a more productive America

The nature of industrial competition is changing rapidly and new challenges are arising. This must be taken into account to improve the industrial practice in the future. The Commission did identify three major and pervasive long-term trends with broad implications for the productive performance of tomorrow's firms:

- economic activity will continue to become more international;
- markets for consumer goods and intermediate goods are becoming more sophisticated partly because of internationalization and partly because of rising incomes around the world;
- the rapid pace of technological change will continue.

The Commission concludes the study with five imperatives for a more productive America that must be adapted by industry, labor, government and the educational community:

- focus on the new fundamentals of manufacturing;
- cultivate a new economic citizenship in the work force;
- blend cooperation and individualism;
- learn to live in the world economy;
- provide for the future.

The study also outlined own strategies for productivity improvement for industry, labor/employer organizations, government and universities.

Focusing on the new fundamentals of manufacturing
The US industry needs to make a major new commitment to technical and organizational excellence in manufacturing after years of relative inattention. Focus on fundamentals in manufacturing has several subgoals:

- establish new measures of productive performance that include such indicators of long-term performance as quality, productivity and speed;
- focus on the effective use of technology in manufacturing including integration of technology in the business strategy;
- embrace product customization and production flexibility;
- innovate in production processes and not only in the front end of the life cycle, to improve the productivity.

Cultivating a new economic citizenship in the work force
Education for technological competence is crucial for raising the productivity of American firms. Effective use of new technology will require people to develop themselves and workers will have a larger responsibility for organization of the production processes. Under the new economic citizenship that the study envisions, workers, managers and engineers will be continually and broadly trained, masters of their technology, in control of their work environment and involved in shaping their firm's objectives. The future will entail improved relationships among companies, employees and technology and learning will acquire greater importance. To obtain this kind of economic citizenship, the study points out three principal subgoals:

- learning about work and at work will include that humans cannot be replaced by automation, but must be continuously trained to use the technology to improve their job;
- increasing employee breadth, responsibility and involvement;
- providing greater employment stability and new rewards based on people participation in the enterprise.

Blending cooperation and individualism
Americans have traditionally emphasized individualism. The challenge ahead is not to suppress individualism in favor of increased cooperation, but rather to combine these two attitudes into a unique mixture that is economically stronger than either extreme. The study discusses four ways for achieving this:

- organize for both cooperation and individualism which will involve flatter organizational structures that invite more communication and cooperation and a culture that rewards both cooperation and individualism;
- promote better intra- and interfirm relations;
- expand partnerships;
- strengthen cooperation between labor and management.

Learning to live in the world economy
To compete successfully in a world that is becoming more international and more competitive, Americans must also expand their outlook beyond their own boundaries. The following subgoals that the Commission suggests that will help foster America's international orientation:

- understand foreign languages, cultures and practices;
- shop internationally to improve the knowledge about conditions abroad;
- enhance distribution and service;
- develop internationally conscious policies.

Providing for the future
There are four major kinds of investments that the study points out as being necessary to achieve a higher level of productive performance:

- invest in basic education and technical literacy;
- develop long-term business strategies and reward more investments that secure long-term profits;
- establish policies that stimulate productive investments;
- invest in infrastructure for productive performance.

2.3 TOPP - A PRODUCTIVITY PROGRAM FOR MANUFACTURING INDUSTRIES

2.3.1 The TOPP program

The TOPP program focuses on improvement of productivity and competitiveness in the Norwegian mechanical and electromechanical industries. The overall goal of TOPP is to:

'Focus on the total productivity for the whole enterprise and stimulate an industrial climate that improves competitiveness'

Important key objectives and key issues in TOPP are:

- time
- quality
- flexibility
- total cost.

The program involves cooperation between:

- the Federation of Norwegian Engineering Industries;
- the Technical University of Norway (NTH);
- industrial companies.

The TOPP program has the following sub-programs:

1. Developing company productivity and competitiveness study;
2. Implementing industrial improvement actions;
3. Developing knowledge through research projects;
4. Long-term competence development (continued education, master degree, etc).

We will not go into detail about the analysis but only introduce the methodologies developed for this purpose. Methodologies are developed for two different approaches (Moseng and Bredrup, 1993):

- Self-evaluation, carried out by the companies themselves based on a standard questionnaire - phase A
- External audit, performed by expert teams which analyze the companies using indicators and key factors - phase B

2.3.2 Phase A: self-evaluation

The preliminary study of competitiveness is the first activity for the participating companies in TOPP. However, the total TOPP program was the focus of marketing. Commitment to performing this comprehensive preliminary study is a prerequisite for full participation. TOPP offered this phase of the analysis free of charge. The main goals of phase A are:

- Develop a methodology for productivity measurement based on a general questionnaire;
- Initiate an internal discussion about productivity issues in the companies;
- Initiate a process for improvement;
- Give the companies opportunities to compare own results with competitors, both domestic and foreign;
- Collect data from Norwegian industry and international industry as a basis for research;
- Evaluate and select companies for phase B.

It is important to notice that this is a preliminary study of competitiveness. Exact measurement is not of primary importance in this phase, but the initialization of a process for improvement. However, much effort was made to make the questionnaire understandable and to explain the process to maximize the validity of the questionnaire.

The questionnaire was developed in cooperation with four pilot companies. These pilot companies represented totally different types of production technology, size and business areas to ensure successful standardization. The questionnaire is divided in three parts:

Part 1: Facts (24 pages). One representative answer from the company
Part 2: Individual evaluation (6 pages) from 10-20 representative people in the organization
Part 3: 'Expert' evaluation (52 pages). One representative answer from cross-functional groups (1-3 groups)

Part 1: Facts
The facts required are dominated by financial data from accounting and finance over a period of 5 years to be able to analyze partial productivity from an economical point of view. The productivity model requires a split of turnover on different product groups with related price indeces, a division of input goods with related price indeces, price indeces for labor, profit/loss sheet and balance sheet. The other questions are mainly chosen to describe the company for instance through the number of employees, customers, suppliers.

Part 2: Internal survey
A representative group (10-20 depending on the size of the company) of people from all levels and functions within the organization is surveyed. They are asked to give their individual view of performance of all functions and resources within the company, and to answer vital questions about top management, quality, organization, attitudes and so forth. These answers often emphasize problem areas that are not mentioned in the expert group evaluation from the company.

Part 3: Expert group evaluation
This is the most comprehensive part of the questionnaire and covers all important activities and resources within a company. The company is supposed to form interdisciplinary groups with 3-4 people to answer questions from each area. TOPP recommends that the group involve experts as well as internal customers regarding the activity. It's also recommendable that each group has a member from the top management both to achieve top management commitment in the process and to be able to conclude in discussions. All questions are evaluations on a seven-point scale. However, there are three different types of questions for each activity as shown in Figure 2.4.

TOPP offer four types of reports from this study:

1. Individual report based on data from own company
2. Report based on data from all participating companies
3. Report based on data from groups of companies. Criteria for selection of groups could be competitors, products, production technology, etc
4. Results from research. An R&D program is established to analyze the total database to search for success factors, etc.

TOPP developed a report generator in connection to the database to handle a variable number of companies. A flexible and understandable report was a main guideline. The report shows the actual dispersion of answers on the scale instead of statistical variance to make comparison easier for the companies. In addition TOPP provided individual comments to the report based on results from the analyses and experience from other companies. Examples from the first two types of reports are shown in Figure 2.5.

PRODUCTION PLANNING AND CONTROL

Evaluate the following factors with respect to current situation, and give a realistic possibility for improvement by an estimate of the status in two years time. Judge the importance on competitiveness for each factor (L = Little importance, M = Medium importance and G = Great importance).

Very bad 1	2	3	Medium 4	5	6	Very good 7

	Current status	Realistic status in 2 years	Importance	*)M...
a) Master production scheduling	1--2--3--4--5--6--7	1--2--3--4--5--6--7	L--M--G	
b) Quality on the production database	1--2--3--4--5--6--7	1--2--3--4--5--6--7	L--M--G	
c) Planning tools	1--2--3--4--5--6--7	1--2--3--4--5--6--7	L--M--G	

To which extent are the following postulates true for the enterprise:

Very little 1	2	3	Medium 4	5	6	Very much 7

		*)M...
a) Is production planning vulnerable due to dependency of key persons?	1--2--3--4--5--6--7	
b) Can production plans be provided for a long time scale?	1--2--3--4--5--6--7	
c) Are time buffers included in lead-times?	1--2--3--4--5--6--7	

To which extent are the following aspects within this area true:

Very little 1	2	3	Medium 4	5	6	Very much 7

		*)M...
a) Are own goals defined (strategic, economic, staff etc.) for production planning and control?	1--2--3--4--5--6--7	
b) Do the personnel know the mission and objectives of the enterprise?	1--2--3--4--5--6--7	
c) Are the personnel motivated?	1--2--3--4--5--6--7	

*)M Missing value due to: M1 - not available, M2 - to much work, M3 - irrelevant, and M4 - not sufficient background

Figure 2.4 Examples of questions (part III in the questionnaire).

PRODUCTION PLANNING AND CONTROL

	Current Status: Very bad 1	2	3	Medium 4	5	6	Very good 7	Average
Master production scheduling			1					
Quality on the production database				1				
Planning tools		1						

a) Individual report (1 company)

PRODUCTION PLANNING AND CONTROL

	Current Status: Very bad 1	2	3	Medium 4	5	6	Very good 7	Average
Master production scheduling	1		9	9	11	6	1	4.4
Quality on the production database	1	3	9	8	10	5		4.1
Planning tools	2	4	4	12	7	7		4.1

b) Report for group of companies

Figure 2.5 Examples of output from phase A.

2.3.3 Phase B: External audit

The external audit is the main analyzing activity in the TOPP program and should result in action plans. Carrying out phase B means commitment from the top management to participate in the process. The B phase is tailored to each company. TOPP has developed a comprehensive methodology, but only relevant parts are used in each case. The main goals of phase B is:

- Establish a methodology for analyzing all factors which have major influence on productivity and competitiveness in the company;
- Use several viewpoints and perspectives in analyzing competitiveness to ensure that totality is considered;
- Offer a flexible and specialized methodology to a wide range of companies in different industries;
- Develop a methodology that is executed by external consultants and still achieves standardized reports and reporting routines.

TOPP is aware of the difference between performance and results. Success is a difficult matter to quantify and could not be solely expressed by short-term profit. There are many examples of companies with seemingly high competitiveness (excellent products, state-of-the-art production technology, etc) that fail in the market. External factors as for instance currency impact, governmental regulations, international policy and international trends has a major influence on the relative competitiveness of a company.

TOPP developed an analysis model with measurement of productivity and competitiveness on two levels as shown in Figure 2.6:

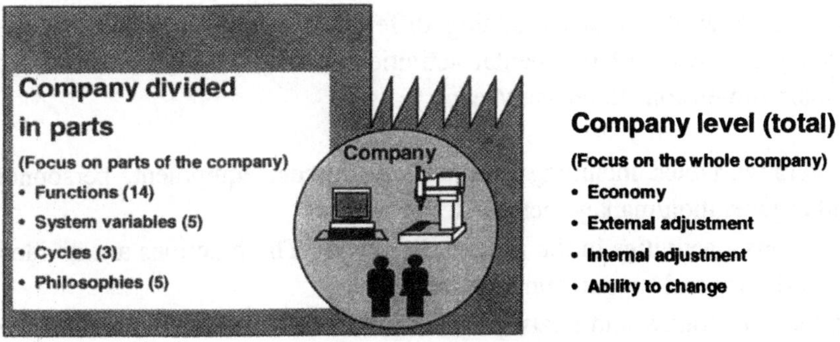

Figure 2.6 The TOPP analysis model

1. Total approach - using performance indicators with focus on the whole enterprise
2. Partial approach - using performance indicators with focus on limited areas in the company

Analysis of competitiveness on the company level
TOPP identified four dimensions to analyze competitiveness on a company level:

- Economy - indicators that describes the financial conditions and potential. An important issue of this analysis is to evaluate the economic capacity for future investments and change.
- External relations - indicators to measure the ability of the company to exploit and take advantage of surrounding conditions. This includes customers, competitors, use of external resources, strategic alliances, etc.
- Internal relations - the ability to use existing internal resources (products, machines, personnel, etc) in a time and cost effective way, and to take care of the internal working environment.
- Ability to change - the ability to foresee and be prepared to meet new trends and quick changes in the environment. The attitude towards change and improvement is an important issue.

Analysis of competitiveness by using the partial approach
This is the traditional view of analyzing companies and focus on internal aspects. Performance is often measured inside each activity or function of the company and results are aggregated to an overall measure. Sub-optimization is a common pitfall when using this method because of the departmental way of analyzing the company. However, the TOPP model avoids these weaknesses by adding different viewpoints to analyze the efficiency of the interdepartmental activities as well. TOPP has identified four main dimensions to be analyzed:

- System variables including products, resources, equipment, personnel and organization/management systems.
- Functions - activities in the product life cycle. The functions are divided into primary and support functions.
- Cycles - to follow and analyze flows between functions. Flow could be of different types (material, information, order, customer, etc). The

objective is to measure interdepartmental efficiency, the infrastructure and cooperation between functions.

- Philosophies - to analyze efficiency and conditions for overall production and management philosophies used in the company.

Planning and performing the analysis

All areas regarding both the 'company level' and 'company divided in parts' are described with a set of indicators. The theoretical model has up to four levels of indicators each containing up to ten sub-indicators. However, only in especially focused areas are all levels of indicators used. This selection of relevance to the company is an important part of the planning process. Every indicator used is given points and they are weighted with respect to competitive importance. The points are summarized to a single value for each level. Figure 2.7 shows a complete example of the indicators describing the Total Quality Management philosophy (TQM). Points are given partly by facts (key factors) and partly by evaluation given by the team members. Demands required to be competitive on the international market are the reference for the evaluation. The scale for phase B is identical to the scale of phase A where:

7 - best practice
4 - on same level as competitors
1 - far behind competitors

The audit is performed by using external expert teams. Normally the teams consist of 2-3 external consultants. TOPP uses a four-step method:

Total Quality Management (TQM)

No.	Description sub indicators	Points	Weight
1	Goals and strategy	3	10
2	Know-how and competence	5	10
3	Organization and management	2	5
4	Information and analysis systems	2	5
5	Strategic quality management	2	15
6	Use of human resources	2	5
7	Quality control and quality improvements	3	5
8	Results of quality work	4	20
9	Customer satisfaction	3	20
10	Cooperation and information	4	5
	Calculated	3.15	100%
	Total valuation	3	

Figure 2.7 TQM indicators (example)

1. Preparation - to present the results from the self-evaluation (phase A), collection of supplementary information and planning of the kick-off meeting.
2. Kick-off meeting - to present information about the methodology (phase B), discuss and give priority to the different areas to be analyzed, and make detailed plans for performing the audit.
3. Perform the analysis - analyze all areas of importance and award points to the indicators.
4. Report - to write the final report and present the results to the company.

3. COMPETITIVENESS AND COMPETITIVE ADVANTAGE

Competitiveness as a comparative measure between companies within an industry is closely related to the presence of a competitive advantage. A major contribution to enlighten competitive advantage has come from studies done by Michael Porter. He describes competitive advantage as follows (Porter, 1985):

> *Competitive advantage grows out of value a firm is able to create for its buyers that exceed the firm's cost of creating it. Value is what buyers are willing to pay, and superior value stems from offering lower price than competitors for equivalent benefits or providing unique benefits that more than offset a higher price. There are two basic types of competitive advantage: cost leadership and differentiation.*

Competitive advantage is a relative term and it depends on the environment each company is surrounded by. Porter's definition of competitive advantage focuses solely on customers and comparative value. This is understandable since exchange of products or service with money is the essence of business. However, this definition could be too narrow to promote an integration between total productivity and competitiveness.

Competitive advantage should not be looked upon as a stable characteristic. A company has to fight to achieve and to sustain competitive advantage. When one company has achieved a competitive advantage the others will start working to even this advantage or even outrun the company. This is the true dynamic of a competitive marketplace.

Development of competitive advantage requires long-term planning and constancy to purpose. It is not just a matter of clever marketing, but a result of integrated effort within the whole organization. This chapter will discuss some important characteristics of competitive advantage.

3.1 COMPETITIVENESS

An increasingly concern for competitiveness in manufacturing companies worldwide leads to some interesting perspectives. Why is everybody so concerned about their national competitiveness? A popular answer on this is of course that the welfare of a country depends on its international competitiveness and hence the ability to export goods with profit. In a business context the interest in competitiveness of each company is not limited to these directly involved, but involves a large number of people and organizations.

It is not surprising, taken into account the difficulties to succeed with a green field investment without sufficient capital and patience, that keeping companies running is a major concern on a national level. Employees are dependent of the availability of jobs, suppliers are dependent on the sales volume of their customers, banks are dependent on the need for capital, the government is dependent on taxes to offer services, etc. The consequences of lacking competitiveness are a chain reaction hurting other parts of commerce and industry. However, the consequences of failures are not limited to creditors, employees, suppliers, but also indirectly hurts the domestic competitors and domestic companies in related industries. Porter (1990) found that internal rivalry, home market demand conditions and existence of related and supporting industry were important explanations for differences in national competitiveness. A loss of a domestic competitor means less internal rivalry. The lesson from Porter is that a single national company in a global industry has small opportunities to survive in the long run. Without support from the other determinants in the 'Porter diamond', long-term success is hard to achieve.

This short discussion clearly shows that many different parties have genuine interest in the competitiveness of a company. A company is surrounded by a number of different stakeholders. The relationship between stakeholder and the company is characterized by an exchange process. Examples could be customers that receive a product and provide money in return, suppliers that receive money and provide components in return and owners providing capital and receiving dividends in return. The stakeholder model shown in Figure 3.1 (Bredrup, 1994a) could be useful to describe these relations

Both the stakeholders and the company have mutual interest in a high level of competitiveness despite the fact that there will always be a discussion of distribution of benefit within the system. However, many of the stakeholders have relationships with more companies or at least the

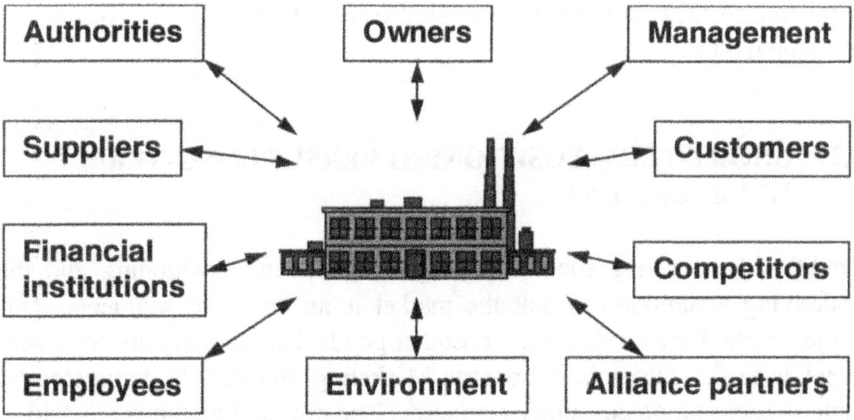

Figure 3.1 The stakeholder model.

possibility of others. This means that the company has to be competitive to attract the best stakeholders. Competitiveness could then be defined as:.

> *Competitiveness is decided by the relative attractiveness toward the different stakeholders. Customers are in an exceptional position among the stakeholders because the only source for payment or benefit to the other stakeholders is the customers.*

Long-term survival and competitiveness are not limited to a battle for the most attractive customers, but include fighting for the most attractive stakeholders as well. Access to the best resources could give an advantage to create value for the customers that exceeds the possibilities of the competitors. It is no controversial statement that competitiveness does not depend solely on own efficiency taken into consideration an average share of purchased parts of around 50% of total product cost (Howell *et al.*, 1987). A company could influence its relative position towards the stakeholders implicating that professional companies will be rewarded. A demanding customer with a high and modern purchasing competence will probably achieve a higher total value than a traditional company. The 'professional' company has stronger buying power than others.

This definition of competitiveness implicates that a company has to operate on several different fronts. Those companies that achieve the best conditions from suppliers, the best conditions from financial institutions, the best alliance partners, the best qualified employees, etc. have a competitive advantage. A consequence of this is that marketing and performance measurement have several different target groups beside the traditional ones.

3.2 SHORT-TERM POSITIONING VERSUS LONG-TERM DEVELOPMENT

Traditional marketing focuses on segmentation and positioning and the underlying assumption is that the market is an ocean of segments. This could be true for a producer of consumer goods, but certainly not for a sub-supplier to the automobile industry. Marketing literature is dominated by authors focusing on consumer marketing. Theories and methods are said to be universal, but this could sometimes be questioned. We have no intention to reject marketing theory, but just point out some pitfalls with traditional marketing for industrial marketers. Our main objection to marketing literature is the isolation of marketing.

Consumer versus organizational buyers

Most mechanical and electromechanical industries have organizational buyers. Their products are often used as a means of production or as components for their customers. Many are sub-suppliers for original equipment manufacturers (OEM). Is it possible to transfer all theories from traditional marketing to these kind of companies? Some of the theory is certainly valid, but major parts have to be modified or supplemented.

A major difference between the consumer and the industrial buyer is that the industrial buyer preferentially is a professional buyer. World class manufacturers are professional and demanding customers and our Norwegian companies aspiring to this title are no exception. However, many companies look at demanding customers as a threat, but the best companies look upon them as assets. Professional customers are a major asset in the fight for productivity improvement. The industrial buyer is presumably more rational than the consumer who could be influenced by good commercials and other marketing. In the consumer market professional marketing could be the most important and perhaps single competitive advantage. This is impossible for an industrial marketer although marketing is extremely important. Nobody will survive in the long run without a clear competitive advantage related to either product or process.

Identification of the real customer is not always easy when dealing with industrial buyers. A large number of people within the buying organization are involved especially for big or long-term agreements. Is the real customer the end-user or the financial department or perhaps the decision maker? Kotler (1988) suggests that a set of different roles in the buying process could be identified within a buying organization. The buying center model is shown in Figure 3.2.

This method is well described for the traditional selling activity, but unfortunately do not cover the whole relationship between an industrial marketer and buyer. Measurement of customer satisfaction and quality perception, procedures for adoption of new requirements, improvement in the interface between the marketer and buyer organization are all important aspects in industrial marketing. This is due to the trend towards long-term relationship between marketer and buyer. Productivity improvement is a condition for long-term obligations. Long-term relationship replaces the competition between a number of alternative suppliers and means new challenges for marketing. A common trend is that marketing is integrated in all activities. Kotler (1988) quotes David Packard of Hewlett-Packard to illustrate this: 'marketing is too important to be left to the marketing department.'

View of the market: an ocean of segments or customer partnership

Segmentation and positioning are central in traditional marketing literature. The majority of literature enlighten the process before the sale is closed. Consumer marketing deals with an ocean of possible customers where segmentation is a tool to try to divide the ocean into the homogeneous parts of the market. This is done to develop efficient market strategies and

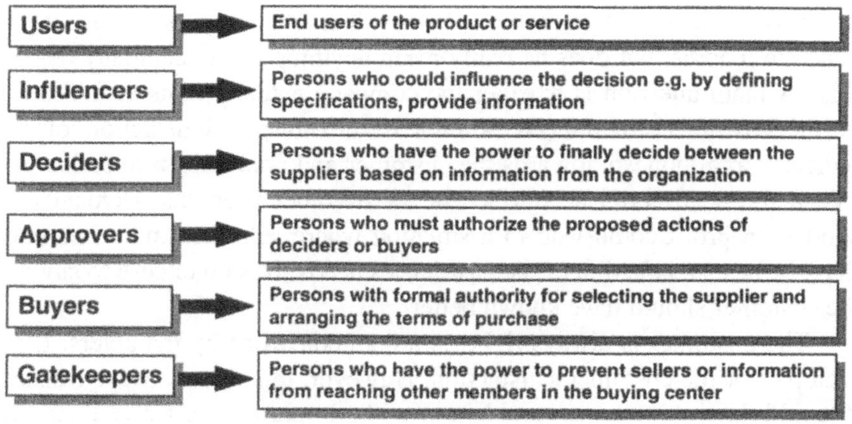

Figure 3.2 The buying center model.

marketing programs. Kotler (1988) divides segmentation variables for consumer markets into four main groups: geographic, demographic, psychographic and behavioral. Wille (1992) divides segmentation for industrial markets into macro- and microsegmentation. Macrosegmentation variables are mainly demographic aspects like industry, size, geographic location, etc. Microsegmentation variables include technology, buyer behavior, purchasing policy, etc. Our main impression is that methods of segmentation are more mature and precise regarding consumer market than for the industrial market.

However, the principle of choosing your customers is important and the segmentation theories could be useful. Many companies do not have their own products, but manufacture customer-owned products. Subcontractors int the electronic industry is an example of this. Consider a subsupplier for the automotive industry. This supplier has a very limited number of possible customers and has to choose the best prospects. Criteria are among others the competitive edge of the customer and the changes of relationship. Segmentation is relevant for industrial marketers as well, but it's only a small part of the total marketing concept. The real important marketing task for most industrial marketers is to develop and maintain a relationship with their customers. Repeat buying is a key phrase for all industrial marketers because of a limited number of customers compared with the consumer market.

Customer orientation

Most companies will claim to be customer oriented, but the TOPP study revealed that practice does not match intentions. However, companies with a customer oriented approach showed a positive correlation to all kinds of business achievement (Bredrup, 1994b). The customer is the means to generate income, but does he really mean anything to the company besides this? A main question is whether the company actually aims at fulfillment of the customer's real need (known or unknown). Satisfaction of this indicates total honesty towards the customer and could mean rejection of a sale if the need could be fulfilled better by other products. Demands for short-term profit could lead to a situation where sales are more important than the best satisfaction of the customer's needs. It is no excuse to say that the customer should have known better

'The customer is always right' is a common word by marketers. It's a vital part of the customer satisfaction. However, this is not always true and it could lead to serious problems for both parts if practiced strictly.

Industrial marketers are assumed to be experts regarding their products. It's their duty to analyze and inform about consequences of new requirements.

A comparison of American, Japanese and German manufacturers done by Hayes and Wheelwright (1984) indicates different views of customer orientation. Whereas most American companies were eager to fulfill every requirement from customers, Japanese and German manufacturers questioned whether the requirements really improved the value for the customer. If new requirements give no real benefit for the customer then it's better to inform him about it. Somebody is paying the price anyway and this somebody is your customer.

The marketer is fooling himself if he accepts new requirements that do not match the process abilities. If a manufacturer accepts all kinds of special adjustments without charging anything extra this means that customers without special orders are paying too much. Flexibility is a powerful competitive advantage, but it's a dangerous strategy without a process technology that makes it possible. An accounting system that is unable to identify the real price of special adjustments could also obscure any mismatch between market demands/needs and the system ability. In such cases the marketer does not know how much to charge for special adjustments and cannot know whether costs for accepting special orders are higher than income. For isolated occasions this could be accepted, but for an increasing number of special orders this could be dangerous. If there is a demand for flexibility the company has to plan for this. Flexibility is no uniform term and the exact qualities demanded have to be identified. Development horizon is long-term. This is one of many examples that illustrates the dependence between marketing and other activities.

Customer orientation is not always easy when dealing with an industrial buyer. Imagine that the marketing department receives a signal from a person representing the customer. Should the company react on this signal immediately or wait for a verification whether the signal is representative for the customer or not? Who inside the customer's organization should be satisfied and where should customer satisfaction be measured? These are difficult questions with no final answers. Still it is important to find the best possible answers all the time relationships between customer and supplier are growing in importance.

Integration of marketing in all activities

If consumer marketing theory is adapted for an industrial marketer without any modifications it is probably doomed to fail. There is nothing wrong with the theory itself, but the use of every tool has to be adjusted to the environment. The most important objection we have to the traditional marketing approach is that it fails to integrate all parts of the company inside the marketing. This is a major obstacle to achieve real productivity improvements. There has to be a close interplay between marketing and all other value adding activities inside the company to identify weaknesses and improvement potentials.

Improvement of competitiveness is achieved only if it adds value for the customer. To ensure this, marketing has to be integrated in all activities. This is not unique for industrial marketing, but probably more crucial here than in consumer marketing. Unfortunately marketing is often an isolated function within many companies. Analyses of the TOPP database have revealed a lack of integration between quality management and marketing (Bredrup, 1994b). Quality improvement does not seem to be customer driven in the companies within the study due to a lack of knowledge of customer preferences and problems.

Modern management philosophies like Total Quality Management (TQM) and Just-in-Time (JIT) are exponents for integration of marketing in all activities. Everybody in an organization is a marketer according to these philosophies. Both philosophies are powerful strategies to develop and maintain a partnership with the customers.

The first step in any marketing planning (Kotler, 1988) is to collect information trough different analyses. Examples of such analyses could be internal and external environmental analysis (SWOT analysis), analysis of the marketing environment, analysis of markets and buyer behavior and analysis of competitors. Together with measures and forecasts about market size, these analyses make it possible to identify market segments, selecting target markets and developing market positions. Then one has to design marketing strategies and plan marketing programs. With great danger of oversimplifying, we could say that a company should look into the great ocean for market segments that accept the competitiveness and products of the company. The critical point is said to be the right segmentation of the market and right positioning of the company. However, competitive advantage is not found by analyses alone, but is a result of a long-term development. We do not want to underestimate the importance of analyses, but stress the importance of long-term planning to develop competitive advantage.

Development and maintenance of competitive advantage cannot be left to coincidences. It's necessary to integrate all strategies to ensure lasting competitiveness. This requires continuous improvement on all matters of importance for customers. Short-term adjustment to rapidly changing customer requirements have to follow a long-term plan to secure effective use of resources and fulfillment of long-term goals.

3.3 DEVELOPING AND MAINTAINING A LONG-TERM RELATIONSHIP

The marketplace is changing for most industries. Adaptation of modern management philosophies like Total Quality Management (TQM), Time-Based Management (TBM) and Just-in-Time (JIT) has led to a new view of the supplier/customer relation. The traditional game of the win-loose relationship is about to change to a partnership based on a win-win situation.

The need for change in supplier/customer relationship

Traditional supplier/customer relationships have been short-term due to the fear of loosing the benefit of competition. The old procurement paradigm said that without a constant threat of competition a supplier had no incentive to improve. The move away from the old paradigm has been a result of force and evolvement:

- Japanese companies implementing the JIT philosophy have experienced a significant productivity increase in procurement. Long-term relationships have resulted in large productivity gains for supplier and buyer.
- Increasing product complexity and decreasing product life cycle times have forced many companies to integrate the supplier network closer to be able to respond fast enough to changing market demands.
- Purchased material has increased its share of product cost in most companies making competitive suppliers even more important than before.
- As high conformance quality is becoming a matter of course in the competition having low quality suppliers is a dangerous strategy.
- As time and flexibility are becoming important differentiation criteria, having supportive and responsive suppliers becomes a necessity.

These factors will inevitably lead to a change in procurement strategies. Sticking to the old strategies could lead to higher costs and a loss of advantage opportunities.

Consequences of a changing procurement paradigm

A massive change in procurement strategies will influence most industries in industrial marketplaces. It will lead to strategic threats and opportunities for buyers and suppliers:

- Number of suppliers will decrease. Companies leave the multisource strategy and carry out single sourcing for more and more items. A 1985 study of automotive industry showed that, although 70% of the companies had traditionally used multi- or dual sources, only 45% were doing so in 1985 (O'Neal and Bertrand, 1991). In addition companies start looking for system suppliers to get a further decrease in the number of suppliers. This development is also central in the lean production concept described in the MIT study on the future of the automotive industry (Womack *et al.*, 1990).
- Supplier and buyer will develop mutual goals. Traditionally negotiations have been adversial, with both parties trying to maximize their individual benefit. However, companies start to realize that buyer and supplier is in the same boat.
- Relationship will change towards partnership. Companies start looking at supplier's value chain as an extension of the companies owns both regarding product development and production. Analyses of productivity and competitiveness will look at the benefit for the total value system instead of suboptimizing each value chain.
- Agreement length will increase. Mutual trust, confidence, and loyalty are critical as two companies establish a long-term, single-source relationship.
- Communication between supplier and buyer will increase. Traditionally communication has been kept to a minimum. Information on costs, product plans, etc. were kept secret due to a lack of confidence. However, exchange of beneficial communication on all levels will increase to maximize the total productivity.
- Communication will be more direct. As productivity for the whole value system becomes the primary goal, new and more efficient communication channels will be developed. Integration of computer systems and direct communication between for instance workers in the manufacturing departments could be a result.
- Supplier evaluation becomes more common and systematic. In some industries ISO-9000 certification will be enough, but others may have even more demanding requirements. Research on the Norwegian TOPP

database has revealed that ISO-9000 in many cases is used as an alibi for quality and hence is of limited use in supplier evaluation (Bredrup, 1994c). As mutual dependency increases the need for an efficient and continuous evaluation procedure emerge. Evaluation procedures will replace competition as the incentive for continuous improvement. The main purpose is to help identify improvement potentials for the benefit of both, but it will also serve as a constant reminder of the basis for a partnership.

- Supplier development becomes a major strategy to strengthen the supplier network. In especially competitive industries raising and developing the suppliers will be necessary to get and keep competitive suppliers. The ability to attract the best suppliers depends on the attractiveness of the company as a buyer. Competence and willingness and ability to exchange competences are among the important factors.

The consequences for supplier industries could be devastating if they still rely on the old paradigm. O'Neal and Bertrand (1991) have identified criteria for JIT suppliers as:

- More reliable on quality and delivery;
- More helpful on technical assistance and product standardization;
- More versatile in lot sizing;
- Demonstrably committed to their specialties or core business;
- Choosier in selecting their customers and awareness of a better 'fit' between buyer and seller which benefits both parties;
- Equipped to take full advantage of computers and telecommunications.

Future marketing - proving that the company deserves a long-term contract

A winning strategy is obviously to get long-term contracts with attractive buyers. Development towards more system deliveries means more emphasis on strategic alliances and network. An essential strategic issue for a supplier is then to find it's place in a successful value system. Job number one then is to be attractive whether the company is in direct contact with the user or a part of a system network. Marketing is to communicate this attractiveness to the marketplace.

Choosing the best customers becomes a major marketing task in the new paradigm. Due to short-termed contracts the risks related to bad customers have been relatively low. However, the investment and risks increase dramatically as the suppliers change toward a long-termed relationship. A natural consequence of a massive reduction of suppliers in the industry as a whole is that the number of alternative customers decreases. Choosing the right customer to target is then critical for long-term survival. The cost of marketing could increase in the short run, but decrease in the long run. Choosing thr right prospects is then critical for survival. O'Neal and Bertrand (1991) suggest a customer evaluation quite similar to supplier evaluation:

- Sourcing philosophy. Customers that rely on their suppliers to add value are attractive to suppliers. An important criterion is whether the buyer commits to long-term external sourcing.
- Geographic location. Suppliers prefer nearby customers. A close location gives better possibilities for positive dynamics in the 'Porter diamond'. A nearby location gives the supplier a strategic advantage and makes strategies like JIT possible.
- Financial strength and performance. Suppliers should evaluate the financial and market track records of the buyer to ensure picking a possible winner.
- Communication capability. For the partnership to work properly, exchange of information is crucial. Information like research and development results, new-product plans, demand forecasts, cost information, quality levels, etc. are probably sensitive information, but the willingness to share is vital for mutual success.
- Market responsiveness. The customer is one step closer to the end-user and his ability to respond to customers' demand affects the supplier directly. The ability to determine customers' needs and quickly meet those needs are an important criterion for the supplier.
- Compatibility with the supplier. This could include having mutual goals, operational philosophies, committed to the same strategies like TQM, JIT or TBM, same information technology basis, etc.

Conducting a customer evaluation is not an offensive act against the customer. To be able to establish a long-term relationship both parties must know each other intimately to have maximum benefit. It is also important to note that this evaluation is no event just to pick out the right customer, but a process. Building a partnership is a long-term process that is a result of a growing confidence in the supplier.

Marketing has to be innovative and integrated with all processes within the company to identify and communicate competitive advantage and added value for the customer. According to the buying center model, advantage and value have to be identified and communicated to the different role figures. There are numerous possible sources for added value and competitive advantage. An A.T. Kearney study suggests improving quality and productivity in the logistics processes as a major source for customer satisfaction breakthroughs (Byrne and Markham, 1991).

A major issue is to try and visualize the value or competitive advantage. This will influence the way marketing is done and totally change performance measurement. It will probably also influence the way cost accounting is done to ensure a supportive financial management.

Performance measurement must change from almost only satisfying the need of the owners to be more externally focused. This is necessary both to identify the most important improvement potentials and to support the marketing to maintain the relationship to the customer. Improving this relationship means recording and communicating improvement. Identifying the right measures is an important marketing activity.

Continuous improvement is the basis for a long-term partnership. Recording and documentation of progress and actions are important to increase the confidence between the partners. Different strategies, methods and tools to improve productivity will be discussed throughout this book. They will be helpful in establishing and maintaining a long-term relationship between buyer and supplier.

3.4 HOW TO REVEAL THE CUSTOMER'S TOTAL VALUE

In this section we only focus on service aspects including quality and logistics performance. Management focus has changed in every decade as seen in Figure 3.3 (Byrne and Markham, 1991). Creating customer value is the main focus in world-class companies. Byrne and Markham (1991) states that there are two ways of creating value: by meeting customer requirements and by exceeding customer expectations (often by providing better value for the price). However, we need to question whether the price is a suitable parameter to determine value.

Using the value chain to identify competitive advantage

Cost and performance could be examined by the value chain (Porter, 1985). Competitors cost and performances could be estimated or found through

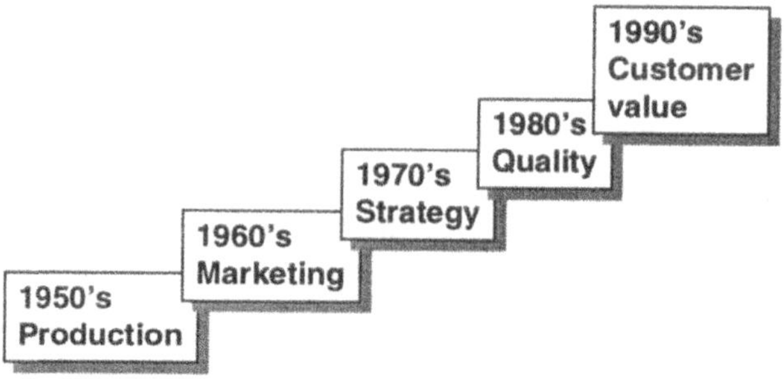

Figure 3.3 Development in management focus over the past decades.

benchmarking. Competitive advantage is determined by the extent to which the company can do better than its competitors.

A company also has to look for competitive advantage beyond its own value chain to justify its position in the value system. Sources for competitive advantage could be found in the value chains of the suppliers, distributors, and ultimately customers. If a company can make the interface to suppliers more efficient or even transfer new methods to a major supplier, savings could improve competitiveness. Suggestions and actions to reduce costs or increase value for the customer might increase the confidence between the partners. The savings could also be split between the partners making the company more profitable.

The value chain is a comprehensive framework for systematically searching for ways to provide superior value to customers (Kotler, 1988). However, the number of possible approaches to achieve competitive advantage and the size of the advantage will vary with type of industry. Figure 3.4 shows the new matrix developed by the Boston Consulting Group in which four types of industries are identified:

Number of approaches to achieve advantage		Size of advantage: Small	Size of advantage: Large
	Many	Fragmented	Specialization
	Few	Stalemate	Volume

Figure 3.4 The new BCG matrix.

- Volume industry. Companies can gain only a few, but rather large, advantages. Includes high volume supplier industries. Profitability is correlated to company size and market share.
- Stalemated industry. Few potential advantages and each of them are small. Includes companies with common and mature technology. Profitability is unrelated to market share.
- Fragmented industry. Companies may face many opportunities for differentiation, but each of them is of small importance. Includes companies with a high share of service included in the product. Profitability is unrelated to market share.
- Specialized industry. Companies have many opportunities for differentiation, and each of them can have a high payoff. Includes companies with specialized products for selected market segments. Profitability is unrelated to size, but is influenced by market share in market segments.

Some advantages may be difficult to imitate because of special conditions or legal restrictions. Minor advantages especially based on new service features are easy to imitate and continuous innovation is required to keep competitors off balance. Advantages based on superior performance, for instance logistics performance, may be more difficult to imitate. A company achieves a cost advantage if competitors imitate at a higher cost.

Revealing the right purchasing price

Cost leadership is one of the competitive strategies identified by Porter (1980). However, cost and price are not directly synonymous and care has to be taken. Competing against low price companies is difficult, but there are several opportunities to beat them with their own weapons. Actual purchasing price is only a part of the total cost of purchasing. According to Scheuing (1989) the main costs affected by the purchasing process are:

- Production costs
- Goods handling costs
- Storage costs
- Capital costs
- Supplier handling costs
- Administrative costs
- Development costs

However, in relation to purchasing these costs are considered indirect and difficult to identify. It is up to the supplier to identify and communicate the advantage and value to the customer. Kaplan (1990) has introduced a model of total cost of ownership of materials. We have illustrated the total purchasing price in Figure 3.5.

According to Byrne and Markham (1991) several companies like Rockwell, Honeywell, Ford, and Motorola have started to develop measurement systems that help focus both parties on the 'true cost' of the supplier's goods. The basis for this approach is to identify all unproductive costs related to doing business with a supplier. Estimates of these costs for each supplier are added to the bid price to develop its 'true cost.' The categories of unproductive costs will vary from company to company depending on factors like type of business, adaptation of management philosophies like JIT and TBM and so forth. A general list could include (Byrne and Markham, 1991):

- Source and receiving inspection
- Rework and returns
- Under- and over-shipment
- Early or late receipt
- In-process rejection
- Paperwork errors
- Failure in the field

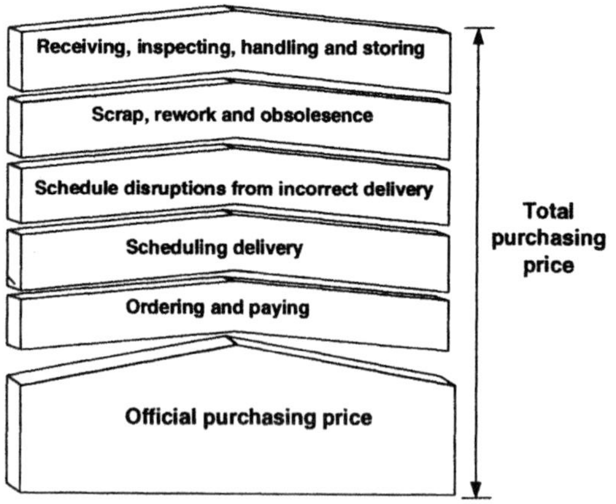

Figure 3.5 Total purchasing price.

Companies calculate a 'true cost multiplier' using the estimate of 'true cost' divided by the bid price. This quota is used in supplier evaluation and companies dedicated to continuous improvement demand continuous reduction of this multiplier. Originally these models and tools were developed for purchasing, but why not use them as input in an innovative process in marketing? Purchasing and marketing are very similar as long as both are operating in an industrial marketplace.

Accepting these models, an apparently low purchasing price might cover high additional cost. These additional costs are easy to get away with in many companies because they do not have a cost accounting system or a performance measurement system capable of isolating these costs without much extra effort. Activity-based costing could perhaps help identify the real costs of a purchasing and highlight the performance of each supplier. In the mean time suppliers have to visualize that their performance adds value compared to other competitors. If a supplier can supply defect-free products, requiring no incoming handling or inspection, in exactly the right quantities, directly to the machine, minutes before the materials are needed, with ordering, invoicing, and payment handled automatically and electronically, a higher price can be justified because of great savings for the customer. Improved quality and logistics performance seem to be keywords for adding value. Steudel and Desruelle (1992) emphasize quality, delivery and cost as the new criteria for purchasing

An A.T. Kearney study (Byrne and Markham, 1991) revealed that 'gaps' between suppliers and customers inhibit quality improvement in logistics for many companies. The suppliers did not understand the customer's service requirements well enough. They rely on sales force input, competitors' performance levels, or customer complaint level instead of asking the customers directly. Another important finding was that only 10% of the companies have the infrastructure needed to support a customer satisfaction strategy. These findings emphasize the need of being customer driven. A customer survey aimed at finding the customer opinion regarding the need for and value of additional service is necessary to close the gap revealed above. The survey must be focused on recording the customer voice and not to justify beliefs, because it is actually the customer who determines value added. However, there exists a difficult balance for all innovative companies and the development of additional services is no exception. The customer may not be able to determine whether a service adds value or not before it is tested.

Visualizing the added value

A marketer cannot trust anybody else to communicate performance to buyers and potential buyers. Visualizing the performance is a major problem and an essential marketing task. This is an ongoing process and not an event aimed at winning new customers. Very simplified we could divide the process in four steps:

- *Identify critical performance criteria.* These criteria must be customer focused. Some kind of a market survey is required to record the customers' voice. Valuable input could come from focusing on the 'true purchasing price', but this could involve education of the buyer. Important criteria are likely to be found among quality and logistics performance. It should be noted that internally-driven service innovations can be adapted as long as they are adapted by the market. This risk is common for all development.
- *Performance measurement.* This includes development of indicators to support the critical performance criteria.
- *Recording development in performance.* This includes registration of measurements over time, collection of information from other sources, generating periodic reports from information systems and the accounting system, systematizing all data and develop reports.
- *Customized communication.* This could include development of customer-specific reports and perhaps even feedback to each role figure in the buying center. Scorecards of the actual performance provided from the supplier to each customer is one possibility. Turning round scorecards and audits traditionally used by buyers to evaluate performance of suppliers is another opportunity. This communication could include all aspects of value-added benefits for the customer. Progress in implementation of new management philosophies like TQM, JIT, and TBM could be an example of information of interest for the customer.

4. BACKGROUND FOR PERFORMANCE MANAGEMENT

4.1 THE SCENE FOR FUTURE MANUFACTURING EXCELLENCE

Nobody has the fortune to know exactly what the future will bring. Countless attempts are made to foresee the future, but the number of different opinions is so big that it is difficult to trust anything. Some common characteristics could be drawn by looking at the most popular predictions. The business challenges and climate today is forcing companies to rethink their business approaches. The 1992 US Manufacturing Futures Survey (Kim and Miller, 1993) revealed the following anticipations among the managers regarding changes in business environment (percentage of respondents that mentioned the issue):

- Increasingly globalized market competition, and cooperation (37%)
- More focus on customer's expectations for quality and time (24%)
- Changing nature of work force: their tasks, attitude, expectations, and capabilities (19%)
- Increasing concerns and regulation for environmental issues (13%)
- Declining or non-growing domestic market (12%)
- Rapid change in technology and shortened product life cycle (10%)
- Increased level of competition (9%)
- Information available more quickly and with broader scope (6%)

A clear trend from this survey is that competition will increase due to more globalization and more demanding customers. As customers are exposed to better products and services their expectations for better quality, service and value will increase. Higher expectations have to be met by improved performance to obtain customer satisfaction.

A dominating paradigm in the Western industry has been the trade-off between speed, flexibility, quality and cost. This belief in trade-off as inevitable is a central part of the mass production mentality (Gilmore,

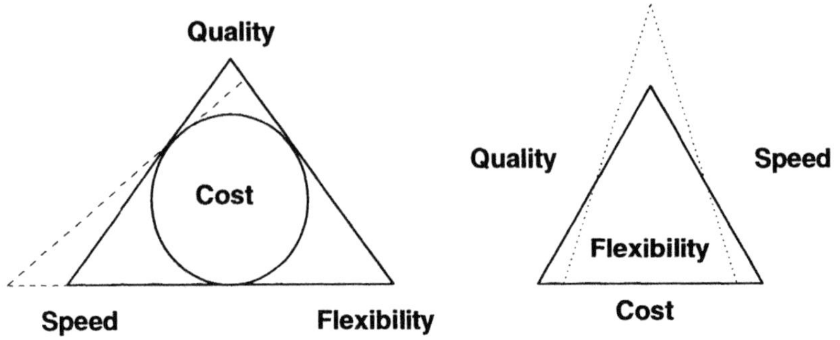

Figure 4.1 Change from trade-off to mass customization

1993). The anticipation's for the future indicate expectations for improvements in all dimensions.

Figure 4.1 visualizes the world of trade-off. According to this, improvements in one performance dimension will inevitable lead to higher cost or to a decrease in the other dimensions as illustrated by the dashed line. Cost and productivity improvement were the major focus. The industrial revolution in Japan has forced western companies to rethink their performance. Implementation of total quality control, Just-in-Time Production etc. has broken the trade-off paradigm and created the world of mass customization with significant improvements in all dimensions as illustrated in Figure 4.1.

Several international studies have looked upon what managers think will be the competitive priorities of the future. Table 4.1 shows results from the *Manufacturing Futures Survey* (Kim and Miller, 1992). A striking result is that price competition is not among the highest priorities in Europe and Japan. Quality and dependable deliveries are among the most important aspects. A 1992 survey gave identical priorities for the US companies.

New challenges have to be met by new tools and methods. Several strategies and methods are available for managers today. However, they provide no quick fix on the problems they are facing. Successful implementation of these strategies involves a radical re-thinking of how

Table 4.1 Competitive capabilities for 1990-95.

Rank	*Europe*	*Japan*	*USA*
1	Conformance quality	Reliable Products	Conformance quality
2	Dependable delivery	Dependable delivery	Dependable delivery
3	Reliable products	Rapid design changes	Reliable products
4	High performance	Conformance quality	High performance
5	Fast delivery	Product customization	Price competition

management is performed. Top management support is not enough anymore. Action is required to change the way management is performed. The old productivity paradigm is still dominating management attitudes in most companies. Major strategies have traditionally been:

- Scale of economies
- Specialization of activities
- Substitution of capital for labor

Specialization is a common denominator with the old paradigm. Activities within administration have often been specialized like quality control, planning, maintenance, etc. A substitution of capital for labor is also a part of this specialization paradigm, because physical capital is usually special in use. Automation often implies increasing task specialization. Increased specialization is assumed to give scale effects. However, Istvan (1992) points out that there are two reasons why capital scale-derived productivity becomes difficult in our mature industrial economics:

- Many processes have reached physical or control limits. Big leaps in technology development become more rare.
- Many processes reach economic limits well before physical limits, which limit the possibility to get scale advantages.

Today's manufacturers experience a pressure for more customization that will lead to more complexity. Gilmore (1993) expects a change from mass production to mass customization. Complexity is traditionally a drag on productivity and will limit the opportunities for scale- and experience-derived productivity improvements. The product life cycle and hence the possible payback time has decreased dramatically in most industries. New strategies and approaches have several common characteristics and focus:

- Customer satisfaction
- Business processes
- Continuous improvement
- Interdepartmental improvement
- Teamwork
- Time as a critical success factor

Implementation of new strategies means a total break with the old productivity paradigm. The 1988 US Manufacturing Future Survey (Miller and Roth, 1988) showed that the traditional management approach gives significantly less productivity improvement than time-focused management. Focus has to move from:

- Accounting to accountability
- Departments to processes
- Individuals to teams
- Managers to management
- Internal measures to customer measures
- Generating reports to generating action and results
- Result management to performance management
- Win/lose relationships to partnerships

These transformation processes are the essence of this book. Management has a key role, because we are talking about a total re-engineering of the infrastructure of the company.

4.2 LEADERSHIP IN MANUFACTURING COMPANIES

Management of manufacturing companies in Western countries desperately needs to be transformed and adjusted to the new competitive environment. New strategies are available and attempts have been made in several companies to implement them. However, measurement systems and management methods often have remained unchanged.

Deming (1986) is one of the toughest critics of Western management methods and philosophies, but he is not alone. Transformation of management is not easy, because there are several 'diseases' inherited in our management culture. These diseases could turn out to be deadly for many companies in the near future. Deming has introduced the five deadly diseases as:

1. Management by numbers
2. Performance appraisal by the numbers
3. Lack of constancy to purpose
4. Emphasis on short-term profits
5. Mobility of management

Nobody really argues against the fact that companies depend on the bottom line. Results should not be the only focus, as Deming says it: 'Manage your firm solely by the numbers, and in five years you will have neither numbers nor a firm to worry about'. But results are the output of a set of processes. Long-term success depends on continuous improvement of the processes. However, the relationships between process improvements and results may be very complex. In most cases it is impossible to translate operational improvements like improved product quality, improved delivery precision and improved flexibility into financial numbers. Financial numbers have traditionally been the dominating performance indicators. Deming states the most important figures often are unknown or unknowable like for instance:

- Future value of a satisfied customer
- Free publicity of a satisfied customer
- Future value of enhanced teamwork
- Improved quality for issues with no specifications
- Improved attitude towards change within the organization

Financial figures are retrospective of nature and provide feedback too late to be a management tool for improvement. Todays management style is often characterized by a collection of figures instead of management of the process. Transformation of management style from managers to management is emphasized as a major challenge for the future.

Deming's diseases have origin in the short-termed financial approach to management. Evaluation of performance, merit rating or annual review for each individual is part of this thinking. Individual rating could inhibit teamwork and encourage suboptimization. The career pattern in many countries implicates rapid change of job either externally or internally. This could encourage managers to focus on short-term results. Results from long-term projects are not available until they have left their position.

Despite the fact that Management by Objectives (MBO) is said to be an almost expelled or outdated philosophy, it is still alive and influencing the management system of most companies. MBO is described as:

> *MBO is a process in which manager and subordinate sit down at the beginning of each performance period and agree upon individual job goals that subsequently serve as a basis for personal performance assessment (Lynch and Cross, 1991).*

> *In addition MBO is a system that begins with the hoped-for outputs, and measures the quality of effort and the effectiveness of resource utilization against these hoped-for outputs (Ordione, 1979).*

The budget process of many companies follows the same pattern, although it may not be called MBO. Hoped-for output is defined in financial terms and the required consequences are broken down in financial budgets for each department. These systems provide no knowledge of factors influencing income and provide no indications of how to achieve improvements stated in the budget. Many companies rely on a system with several independent variables and still expecting mathematic relations to exist. Financial measures like for instance Return on Total Capital or Return on Investment are equally dependent on income and cost. Still many companies overlook the income or actually misuse unexpected variances in income to explain differences from budget.

Often there exist no map for the lower managers to achieve the hoped-for outputs. Let us assume that the productivity increase is set to 5% next year. Why is it exactly 5% and not 7%? If there exists no map for the improvement then it makes no difference what the target in the budget is. In fact those targets are often self-realizing, which often mean that a higher performance is not achieved. Lynch and Cross (1991) have identified five major problems with MBO:

- Fosters internal competition rather than teamwork because it is a mechanism for assessing individual or departmental contributions at review time.
- Goals are set by senior management that may or may not be strategy related. MBO goals often ignore the customer and are based on past performance.
- Focuses on the manager-subordinate (or departmental structure) within the company's vertical structure, rather than on the horizontal work flow independent of organizational boundaries.
- Objectives once set, tend to become fixed throughout the organization. Adapting to changes in external environment is often difficult.
- Measures tend to be one-dimensional, often financially oriented.

Many performance measurement systems available have the same weaknesses although few companies are willing to admit having a MBO system. These companies could be diagnosed as suffering from Deming's deadly disease: management by numbers.

MBO is said to have gone through three phases over the past twenty years (Sink and Tuttle, 1989):

1. Performance appraisal. Emphasis was placed on jointly developing criteria and standards for individuals in a given job.
2. Planning and control method.
3. Method for managing productivity by objectives.

This latter use is still widespread despite heavy critique from many authors, among them Deming (1988). This critique is not regarding use of objectives, but the unbalanced focus on easily available financial measures.

Managers also need to rethink how they spend their time. Our request on leadership for improvement is often met by arguments of little time. However, management time could be divided in three groups as suggested by Sink and Tuttle (1989) in Figure 4.2.

Division in three equal sized slices is not realistic for most companies. Cater for crisis is probably the dominating task for most managers. Many companies have an organizational culture where the 'firefighters' are the heroes. Their style is vigorous and result-oriented, which match the Western management ideals perfectly. However, these 'firefighters' often represent the short-termed focus that are about to kill many companies. Short-term solutions in an environment characterized by rapid change may at first sight look appropriate. Short-term solutions on crisis could lock the company in more or less incidental solutions that turn out to be inefficient and expensive. Decisions are made fast with limited information. It also

Figure 4.2 Use of management time

limits the ability to comply with the long-term strategy. A more cost-effective strategy to meet a rapidly changing environment is to develop the ability to change. Activities for building the business are often the looser in the struggle for attention and resources. This has turned out to be an evil circle in many companies, because the crises are increasing due to low attention on business improvement.

Japanese management style differs radically from the Western regarding time spent on improvement (Imai, 1986). Figure 4.3 shows this difference on the various organizational levels in Western and Japanese companies.

Figure 4.3 suggests that there is little room for continuous improvement in Western companies, because most attention is given to maintenance. Strategic leaps through innovation are focused as the means to improve.

The 1992 *US Manufacturing Futures Survey* gave interesting figures describing how plant managers spent their time as shown in Figure 4.4. It shows that 50% of the time is spent with supervision and internal coordination, whereas little time is spend on integration with marketing, customers, suppliers and R&D. The trend shows that it is about to change in most companies. Emphasis is moved from supervision to integration.

4.3 THE MISMATCH BETWEEN ACCOUNTING SYSTEMS AND CHANGING ENVIRONMENT

During the first half of this century mass production of standardized products was the rule. Management focused mainly on maximizing output produced by direct labor and machines operated by the workers, because this was the dominating part of product cost. Competitiveness was directly dependent on managing these costs. Accounting systems that set standards and closely tracked the productivity of individual workers and machine

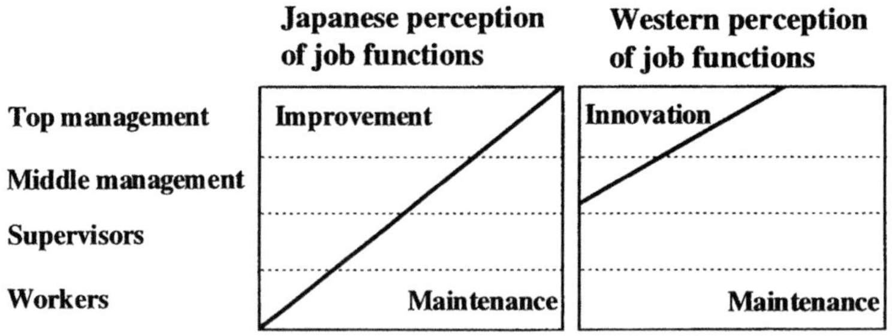

Figure 4.3 Perceptions of job functions.

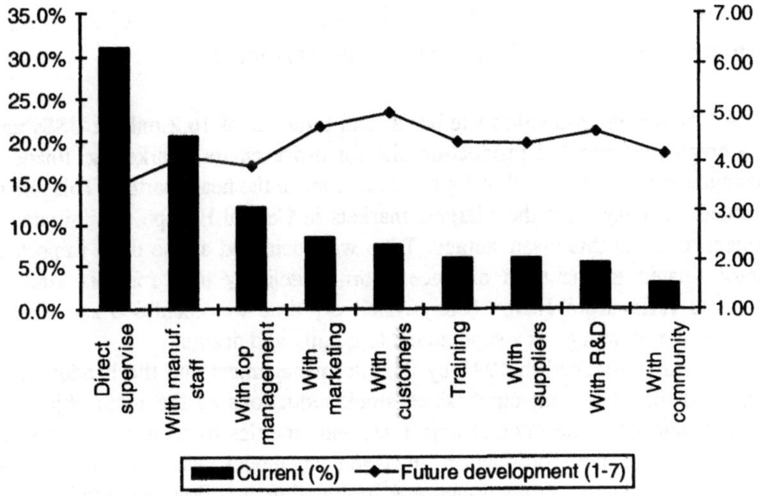

Figure 4.4 Dispersion of plant manager's time.

efficiency were developed to cope with this environment. Ordione (1979) describes the integration of standards into General Motors accounting system back in 1925.

Traditional accounting systems are under heavy fire for not providing relevant and fast enough feedback and generating misleading product cost data. Lynch and Cross (1991) link the problems of accounting to outdated standard costing systems. A workshop at Harvard Business School in 1989 (Kaplan, 1990) concluded that existing systems for cost and performance measurement had lost relevance. They provide little motivation to support companies' attempts to incorporate TQM, JIT, and Flexible Manufacturing Systems into ongoing, continuous improvement activities. This is frightening because many companies invest heavily in such strategies.

Studies have revealed several examples of significant operational improvements in defect rates, setup times, throughput times, etc. that goes unrecognized in the financial statements of the companies. These findings have been well documented by Johnson and Kaplan (1987), Kaplan (1990) and Lynch and Cross (1991).

Financial management consists of three main activities of special interest for performance improvement:

- Investment justification
- Product cost
- Performance measurement

The following example could visualize some of the difficulties related to performance and support from financial management.

A Norwegian production site has a total turnover of 16.2 million US$s and 125 employees and is a production site for office chairs. Marketing, financial management and product development are done at the headquarters. This site is situated far away from their largest markets in Central Europe and initiated a project to offset this disadvantage. Time was identified as the most important factor related to the term distance. Normal delivery time for this kind of products varies from 14 to 30 days. Delivery time was identified as a future competitive advantage as a supplement to quality and design.

JIT was introduced in 1991 by the site management and the headquarters approved the plans, but emphasized stock reductions as the main objective. They developed some internal objectives and metrics to monitor progress in the project. A new production management model was introduced, that changed the layout from functional orientation to flow orientation and implemented a KANBAN system. The most important supplier was integrated as a part of the production system and single sourcing was implemented on transportation services.

Among other results the project made it possible to abolish storage of finished goods and they achieved a guaranteed delivery time within Europe of minimum 1 day to maximum 5 days (different zones). Their changeability and their flexibility so meet unforeseen changes in demand were dramatically improved. However, they experienced an increase of total stock, because of an increase in raw material and work in process. The accounting department was then unable to record any improvement with the JIT project and criticized it heavily. The key question is what their figures would have been without the project, because our evaluation showed changed conditions.

In the project period they launched a totally new product line without drawing back the old line as planned. During the project they discovered that they where able to move the whole production from a site abroad, because of released space. Due to the reduction in delivery time and better coordination of transportation, regional stocks became unnecessary and they received obsolete parts from these stocks.

On the other hand the purchasing department started to source parts from Eastern Europe with low price, poor quality and long delivery time. This resulted in savings in purchasing, but led to increased costs in production and made difficulties for the JIT project. The design department still allows new changes and new details from each designer (3-4 different designers) and have no request for standardization and reuse of parts.

All this resulted in an increase in the number of articles and a temporary increase in tied capital. However, they did manage to handle the new articles and products with the same space. Most of these results are difficult to record with the traditional accounting system. For an outsider it was unbelievable that their evaluation of the project did not take into account the strengthening of the competitiveness and the reduction of future costs.

Investment justification

Several methods are available for evaluation techniques for investment projects. An investment involves a sacrifice of something now for the prospect of something later. It involves time and risk. Evaluation of investment projects is necessary to find out whether the benefit exceeds the sacrifice. The best known methods are:

- Net Present Value (NPV) - the sum of annual cash flows discounted to a particular time to zero. The annual cash flow is the difference between cash inflows and outflows for a particular time period.
- Internal Rate of Return (IRR) - the discount rate at which the Return NPV equals zero
- Payback Period (PBP) - the time interval between the start of sales and the point at which the total cash flow becomes positive.

Park and Sharp-Bette (1990) emphasize that NPV and IRR are theoretically correct, but the practitioners often prefer PBP. The reason why NPV and IRR are more correct is the fact that 'money has a time value'

- Money has a potential earning power. An amount of money available at different points in time have different values.
- Money has a time value. A user may have a different utility of consumption at different times
- Buying power changes through time (inflation)

Several surveys are conducted to find the most popular methods in the industry. A recent survey (Remer *et al.*, 1993) revealed in 33 American Fortune 500 companies the Net Present Value (NPV) method as the most widespread project evaluation technique currently used in US. A comparison with a similar 1978 survey shows a shift from Internal Rate of Return (IRR) to Net Present Value (NPV). However, most companies are using both methods. Payback Period (PBP) has lost popularity, but it is still used by 64% of the surveyed companies. An average weighting was assigned to each method by the companies based on the degree of use. The average weighting of the NPV method was 41%, weighting of IRR was 35% and PBP only 12%.

A 1987 survey by CAM-I with 350 valid responses (Howell *et al.*, 1987) showed a close race between PBP, NPV and IRR. 71% of the respondents used payback period whereas 69% used IRR and 64% used

NPV. These numbers indicate that the respondents are using several methods, but this survey has not weighted the different methods to reveal the real dispersion among them.

A study of Japanese appraisals methods for investments reveals that 65% uses the payback method whereas only 6% use IRR and 10% the NPV method (Sakurai, 1990). The main reason for the widespread use of the payback method is the rapidly changing environment industrial companies are facing. Fast payback is necessary today, given the short life cycle of many products and production technology. Japanese companies said that since sales forecast methods are unreliable it makes no sense to use sophisticated financial methods to calculate future profit (Sakurai, 1990). Sakurai found that 84% of the Japanese companies were using a payback period of five years or less, which matched the CAM-I survey.

However, several authors emphasize problems with these traditional methods. Many companies experience problems with justification of new strategies and tools. It could be questioned whether these methods are able to support transformation of a company. Lynch & Cross (1991) states two major problems with investment justification related to accounting methods:

- Projects are viewed from a local viewpoint and the impact on the local operation or function is evaluated. Downstream and upstream concerns are seldom evaluated properly.
- Finance departments typically focus on discounted cash-flow analyses. Traditional cost savings in terms of labor and space are usually quantified in detail whereas little attention is given to the more intangible benefits like quality, cycle time and delivery.

This shows that traditional or recommended financial methods may fail to support investments of providing benefit of significant importance to competitiveness. They may overlook cross-functional gains and less tangible benefits like quality, cycle time and delivery. A common problem with justification of modern approaches is the requirement for translation of improvements into financial terms. King (1993) emphasizes that savings could be of different kinds. He split savings into two groups:

- Green savings - real cash savings visible in accounting systems as for instance reduced purchasing prices, lay offs, reduced use of overtime and reduced use of materials. Real savings could be reinvested outside the company.

- Blue savings - indirect savings not necessarily visible in accounting like for instance reduced need for space, reduced lead-times, changed organizational culture, improved delivery precision and improved product quality for the customer. These are internal savings or savings that are beneficial for the customers. They are not real cash savings that could be reinvested outside the company.

Investments in modern philosophies and methods like for instance TQM and JIT often result in mainly blue savings. Green savings are often difficult to achieve directly if performance is not very poor before the project starts. This is especially true for small- and medium-sized companies. Saving half a person is not a saving unless half his salary is removed from the budget. The same argument could be used against idle space. On the other hand we face problems with unknowable numbers like how much a satisfied customer is worth in the next period. Another problem could be how much a change-oriented organizational culture will save costs in the future. These examples indicate that the input for the evaluation methods is not valid for most improvement projects. This will also be the case for investments in production technology where increased flexibility, reduced lead time, etc. are the expected benefits.

However, 80% of the respondents in the American study indicate that operational performance improvements, such as quality, technological capability, and delivery, are 'extremely' or 'very' important criteria in evaluating investments (Howell, *et al.*, 1987). A majority of the respondents said that these indicators were only considered qualitatively, meaning that they were not quantified in dollars. Other major qualitative indicators for justification of investments were improved competitive position, constancy with business strategy, improved delivery and service and improved product quality/reliability.

The CAM-I study showed that almost 60% of the respondents felt that their methods needed improvement. Most important suggestions for improvements were quantification of the financial impact of not making an investment, post-investment evaluations and a stronger long-term orientation.

Traditional evaluation methods for investments have severe problems with the ability to support transformation projects towards a more adaptive organization. These methods could easily be used against projects like implementation of Just-in-Time and Total Quality Management. The result is probably given in advance as long as financial consequences are so difficult to assess and the payback time is long. A study by Jewkes and

Power (1993) revealed that investments in JIT do not reach a positive Net Profit Value the first 4-5 years.

Analyses of the TOPP database revealed that investments in modern manufacturing technology can not be solely justified by traditional financial evaluation methods due to problems with transforming benefit into cash savings (Bredrup *et al.*, 1994). Justification has to focus on performance improvement.

Economic evaluation methods are a necessary part of a more balanced approach, but should not be used alone. Input to the methods are in most cases based on the past and more rapid changes will result in a less predictive capability. It is of no use having a theoretically correct method if the basis for calculation is wrong. Investment justification both pre- and post-evaluation needs to focus on total performance improvement to support development of competitiveness.

Product cost

Product cost is an essential measure because it gives a link between cost and income and influences strategic decisions regarding the product portfolio. Basically it should provide answers on the following questions:

- What product lines are making money?
- What product lines are loosing money?

Standard costing is by far the most widespread cost accounting system (Howell *et al.*, 1987). They give the following explanation of standard cost system:

> *A standard cost accounting system develops expected costs for material, labor, and overhead and then uses these standards as a basis for flowing the costs through the various stages of the manufacturing process. The standard costs are compared to actual costs, and any differences and variances are analyzed.*

Direct labor was the dominating factor in product cost when the standard costing system was developed. Overheads played a minor role and allocation was easy. Today the cost structure is totally different as shown in Figure 4.5 (Howell *et al.*, 1987).

Overhead stands for 32% of total product cost (Howell *et al.*, 1987). Overhead is expected to increase rather than decrease due to further reduction in direct labor. Allocation of overhead is no longer a simple task

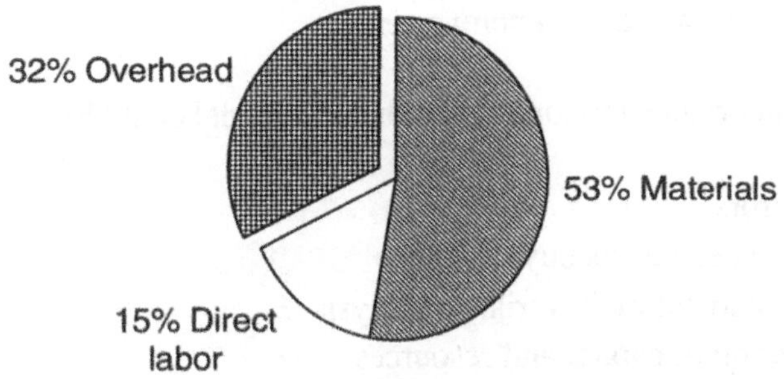

Figure 4.5 Cost structure in advanced manufacturing industry.

in a complex company, and the magnitude of overhead costs questions the reliability of product cost calculations. Calculation of product cost is traditionally based on the simple model:

Product cost = direct materials + direct labor + overhead

Overhead allocation is the critical part of this model. The CAM-I study (Howell *et al.*, 1987) revealed that one-third of the companies used a single overhead rate for the whole company. Another third had separate overheads for groups of work centres, while almost 40% have them for each work centre. A condition for a useful allocation rate is homogeneity. However, most companies have a complexity that makes development of allocation rates a difficult task. It is incredible that management can have any confidence in the product cost calculations when a single overhead rate is being applied universally across the whole business.

The overhead rate is often calculated based on direct labor that often is in favor for less work-consuming products. Several studies have documented severe shortcomings of the cost model, including Lynch and Cross (1991), Sink and Tuttle (1989), Johnson and Kaplan (1987) and Howell *et al.* (1987). Imagine what consequences wrong cost models have on decisions on product pricing, product sourcing, product mix, and responses to rival products. There are several reasons for the loss of relevance:

- Dramatic increase in overhead
- Increasing complexity both on products and process
- Rapidly changing technology regarding product and process
- Increasing customization and number of variants
- Large variances in product mix

- Wish for cross-functional improvements

The most important shortcomings of the cost model could be:

- Pricing errors
- Improper make versus buy decisions
- Irrelevant and untimely variance analysis
- Misallocation of capital and resources
- Non-productive, expensive vouching
- Misidentification of cause and effect relationships
- Improper design incentives

These weaknesses could lead to wrong decisions concerning productivity and competitiveness. Activity-based costing (ABC) is advancing as an alternative to the traditional cost model.

ABC was introduced by Johnson and Kaplan (1987). ABC's objective is to distribute the overhead costs to each product with the aid of sundry cost drivers. Traditional costing systems assume that products cause costs to be incurred whereas ABC assumes that products incur costs according to the activities that they require. ABC proclaims to be innovations developed recently, but the activity concept was introduced in the 1950's by Erich Kohler (1952). He used the term 'Activity Accounting,'. The difference today is the technological possibility that has emerged through computers. A company-wide activity-based system requires data processing possibilities beyond a manual recording system. The main lesson from ABC is that we must manage activities, not cost. To get costs under control, we have to examine the tasks performed.

King (1993) points out the paradox of cost-cutting programs. Cost reductions and process improvements have to be done simultaneously otherwise expected results will hardly be achieved. Unless a corresponding reduction in the actual work load occurs, cutting staff by 10 percent cannot work. The challenge is to move emphasis from cost management to performance management.

In 1990-91 financial institutions in Norway experienced a total collapse. Thousands of employees were fired to cut costs. However, only small adjustments were done in the processes and the same institutions had to hire people from temporary staff recruitment agencies and consultancy firms. Several of these people were actually their old employees.

Performance measurement

Management and decision making depend heavily on representative real time information of status and development over time. Measurement is important in many ways and the following statements may stress the importance of identifying the right metrics:

- You cannot manage what you cannot measure
- What gets measured gets done
- Measurement influences behavior

Efficiency is the dominating dimension in the measurement system of most companies. Measurement systems are often dominated by financial measures directly available from the accounting system. One of the reasons for the popularity of financial measures is the possibility to aggregate performance from different departments. This was one of the main objectives when standard costing was developed in the 1920's (Ordione, 1979). As a paradox, departmentalization is today one of the real problems with traditional measurement systems.

Performance measurement systems should be able to support continuous improvement. 'Blue savings' are often the dominating result of a continuous improvement process. They could be accumulated over time and transformed into real cash savings or as a reduction of future costs. Reduction of future costs is perhaps the most common result due to many reasons. Lay-offs may have a negative impact on new ideas and simply refraining from employing new people is acceptable for all parties. Idle space could seldom be sold unless a whole building or site becomes idle. However, introduction of new product lines or growth could be possible without investments in new buildings. Performance measurement then becomes essential to:

- Justify investments and effort in the improvement process
- Manage the improvement process and ensure compatibility with strategy
- Manage the transformation process from blue to green savings

Without ability to monitor improvements it is difficult to justify further investments and also to market the project internally. There is no doubt that resistance forces against changes are powerful in most companies. If the measurement system is unable to record improvements from projects, resistance forces will get all the ammunition they will need. Companies have grown so complex that it is difficult to detect and to evaluate improvements.

Measurement to record improvement is actually a condition to be able to exploit small improvements.

Performance measurement plays an important role in productivity improvement. If a company's short- and long-term performance measurement is not consistent with the ultimate objective of the company, the likelihood of success is remote. Traditional performance measures provided by accounting data is concerning sales, net income, return on investment (ROI), actual versus budget performance, labor and machine efficiency, work in process, value of inventory, etc. They give a good picture of the financial health, but do they support improvement of the performance required in the marketplace?

Traditional measures are not enough to direct managers towards the objectives of long-term profitability, flexibility, high quality, reduced throughput times, reduced lead times, employee satisfaction and customer satisfaction. Several of these indicators are measured informally, outside the accounting and operating control system (Howell *et al.*, 1987). However, it is the indicators provided by the finance department that reach the top management and the board of directors and hence influence investment or new project decisions. Case studies in TOPP (Bredrup, 1994a) have shown that a major problem is that evaluation is built into the financial measurement system. Evaluation is done by direct comparison of figures and no corrections are done due to changing conditions. Lower level managers are held accountable for financial measures and their behavior as influenced by them. Several studies have revealed severe shortcomings of financial measurement systems. However, it should be emphasized that the major problem is the automation of evaluation built into the metrics that is causing most problems. This source of error could be due to:

- Volume change
- Mix change
- Conflicting metrics
- Omission of actual improvements

Volume change. Cost per unit produced or processed is a common productivity indicator in a standard cost system. Variances against standards are analyzed carefully and determine the productivity development. Combining this with the allocation of capacity-related costs based on actually processed volume rather than on practical capacity can lead to some very wrong conclusions. As the volume decreases, either as a consequence of reduced sales, increased functionality per part, or even due to removal of required rework as quality is improved, the cost per unit metric will increase. Management's conclusion might be that productivity has decreased or cost control has been slackened. Increase in operating efficiency may be offset by decreases in earned hours, due to lowered volume.

An electronic company experienced a decline in production of 12.5%. Since the majority of costs were in the semifixed and fixed categories, which had not declined with the short-term volume decrease, the unit cost rose 14% due to the volume change alone. Reinforcing this trend were activities to certify vendors to avoid acceptance test. The cost per incurred direct hour increased as more vendor parts were certified to be 'free of test'. However, the total cost decreased. Interpretation of the measures could result in a loss of productivity.

Another volume effect could be seen in comparison of different assembly technologies using cost per incurred direct labor hours (DLH) as criteria. Capacity utilization in the introduction phase of new automation technology may be low due to inertia in re-designing parts. The defender naturally has a lower cost per incurred direct labor hour and is preferred. The opposite approach is to take maximum advantage in the new machine to justify the investment. However, both approaches operate with fictional costs and will lead to suboptimization. Machines may become bottlenecks due to these incentives and operational performance could suffer.

Mix change. The life cycle time of most products is decreasing while product complexity is increasing. This makes standard unit cost invalid. Productivity measured as unit cost may decrease despite substantial improvements. This could lead to totally wrong decisions about out sourcing. The effect of mix change is quite obvious but often overlooked.

A manufacturer of cables experienced a major change in mix. The sales volume decreased measured in meters, but remained constant regarding dollars due to changes towards thicker cables. Profit increased, but the measurement system gave a warning due to increased unit price. Imagine that the company has only two products A and B.

Income in year 1 is given by which SP is sales price and SV is sales volume:

$$I_1 = (SP_A \cdot SV_A) + (SP_B \cdot SV_B)$$

Cost in year 1 is then given by where CP is cost and SV is sales volume:

$$C_1 = (PC_A \cdot SV_A) + (PC_B \cdot SV_B)$$

Profit year 1 (P_1) is given as

$$P_1 = I_1 - C_1$$

Cost per unit year 1 (CU_1) is given as

$$CU_1 = \frac{C_1}{(SV_A + SV_B)}$$

A calculation for two years with numbers could illustrate the effects:

Year	SP_A	SV_A	SP_B	SV_B	I	PC_A	SV_A	PC_B	SV_B	C	P
1	7.5	2000	6	5000	45000	6	2000	5	5000	37000	8000
2	7.5	3000	6	3800	45300	6	3000	5	3800	37000	8300

The result of this change in product mix was an improvement regarding profit, but an increase in cost per unit from 5.28 to 5.44. This example is oversimplified. In a real situation the company may have so many variants that calculations for each variant become too expensive. Cost per unit may be calculated only for product groups. In the original case the management evaluated productivity to have fallen.

Calculations of cost per unit become impossible in a situation with extensive customization. Validity could be questioned if calculations of unit cost have to rely on homogeneity inside a product group. A change in product mix will probably give changes in cost per unit caused by the calculations. Cost per unit is often used as a productivity metric but wrong conclusions could easily be drawn.

Conflicting metrics. A common problem with all kinds of metrics is the appearance of conflicting metrics. Departmentalized measurement systems, as the standard cost system is an example of, could lead to conflicting performance metrics. The result is that one department is trying to influence something that will lead to a loss in performance in another department. A classic example is purchasing performance measured with price variance (PPV).

A possible action to improve the price variance is increased order quantity to get lower price, ignoring quality and delivery. This could result in excess inventory, increased carrying costs and overlooking the total price of purchased materials. Purchasing may be evaluated as making good progress in meeting costs, while the actual costs of the policy are shifted onto manufacturing costs.

> A purchasing manager of a company aiming at JIT had developed a total cost model for suppliers. However, he was only held responsible by top management for the price variance. Consequently he had to choose suppliers from Eastern Europe with low cost. The company soon experienced severe quality problems and delivery problems. A delivery delay for a key component caused a 3-day stop in production.

A critical task in development of a performance measurement system is to ensure that conflicting metrics are removed. Measurement influences behavior and use of metrics has to be carefully audited against strategy to ensure supportive behavior within the organization.

> A manufacturer of parts for the automotive industry audited its measurement against their strategic plan during a JIT project. One of their metrics was to ensure that all employees were at work from 8 to 16 hours every day. However, their obligations as a JIT supplier meant that production varied from day to day. This meant that the company had to use overtime some days and pay the people to do nothing other days. They changed to a more flexible system where the employees managed their own working time. The employees are now responsible for smoothing out the capacity with their working hours. This system is to the best of both parties.

A common accusation of traditional accounting methods is that they do not punish overproduction. In fact companies may have a consistent 'policy' of overproduction, causing high inventory levels, risks of obsolescence, lack of parts, etc. All of these problems can be traced back to the financial measures being employed to evaluate the operations. Each machine's individual efficiency may be evaluated, based on direct labor hours earned, estimated as standards for each component. Using the ratio of earned to actual DLH as a productivity indicator, means emphasizing that the machines should be kept busy at all times, even if it means producing way ahead of schedule.

> A manufacturer of agricultural machinery decided to implement JIT in one of their assembly lines. We analyzed their standard cost-based measurement system and found that the improvements were not likely to be recorded by their metrics. They measured productivity of direct labor, but handling and transportation were held outside the standard system. One of the actions in the project was to streamline the material flow and to reduce handling and internal

> transportation to a minimum. This was only possible with a synchronization of speed for each assembly station. Synchronization meant a lowering of speed in some stations resulting in a loss of productivity. The project was not credited for reducing handling and transportation, because this was outside the system and shared with other assembly lines. Saving half a person in this case will not be recorded as an improvement in the accounting system unless the pay by reduced by the same magnitude.

A measurement system that encourages larger batches could result in severe operational problems for a company when the volume varies. An example of such motivation is to exclude setup hours from the earned hours. Larger batches will result in high inventories and will result in shortages in a situation with high capacity utilization.

> A manufacturer of large engines experienced a situation with frequent shortages in a period with high capacity utilization. These shortages resulted in extensive use of extra resources to keep the delivery dead lines. An analysis of this problem revealed that all kinds of parts were missing even very cheap ones. At the same time the value of inventory increased. Despite heavy investments in flexible manufacturing equipment large batches were the rule. This was due to a shortsighted change of production planning and incentives in the measurement system. Setup times were not included in earned hours. In a period with high volume the 'system' promoted a reduction in setups leading to larger batches and hence larger inventory. However, the inventory mix was wrong and frequent shortages were the result.

Sales per employee or manufacturing costs as a percentage of sales is common indicators of performance. Large corporations often use it to compare performance of different sites. Analyses have shown that these metrics could have adverse effects as incentives for a reduction in value added activities. In an efficient production system use of direct labor is synonymous to value added. However, comparison of use of direct labor without taking into account the efficiency could lead to wrong conclusions about productivity

> Number of units produced per employee were used to compare two units within a multinational corporation. However, one of the companies manufactured most parts by themselves whereas the other was a pure assembler. Work in progress was compared in actual dollars for the same units without take the level of value adding into account. The result is a powerful incentive for the manufacturer to out-source more parts.

Omission of actual improvements. The main reason for this omission is that benefit occurs outside the range of the metrics. Improvements in inventory level and corresponding floor space needs may be overlooked because the accounting system may be unable to record them. Cost savings occurring in other units may be omitted like for instance quality improvement, improved delivery precision, reduced delivery time, and reduced lead times. Improved defect rates are measured by the customers and increased value or reduced costs for the customer are not valued in the measurement system. All these improvements are what we earlier defined as 'blue savings'. They could eventually be transformed into direct savings or increased sales, but are not possible to record in a traditional financial-based performance measurement system.

A major problem is that the conditions change during the measurement period. Evaluation is often built into the indicators in a financial measurement system. This means that comparison is done directly between two periods. However, the conditions may have changed leading to an improvement in performance despite unchanged level in performance indicators. Changing conditions could be an increase in number of products, articles and variants within the system. Another reason could be that benefit is distributed before measurement is done.

A manufacturer of equipment for distribution of electricity started a broad-based JIT project in 1992. They are competing mainly on price, quality and delivery time. A significant increase in exports has led to a demand for productivity improvement due to a lower gross margin on the export markets. A change from wholesalers to direct sales has led to more pressure on delivery times in the home market. They also experienced seasonal variances. Main objectives for the project were:

- Reduced stock
- Reduced outstanding claims
- Reduced throughput times
- Reduced delivery times

The JIT project consisted of ten work packages with dedicated project managers for each. Those packages were training, KANBAN (flow orientation), reduction of setup times, quality improvement, maintenance, salary, information technology, measurement system, productivity improvement in administration and sales and improvements in product development (time, cost and quality).

As a part of the project they laid a hardened epoxy floor and changed the layout completely. All stocks (raw material and components) were moved to the shop floor the organization of each shop floor was changed to be more efficient. Standardized boxes and shelves with wheels made the internal material flow more efficient and flexible.

No performance measures except financial where available before the project started which made comparison impossible. However, they started during the project to measure some essential metrics corresponding to the objectives. They did not achieve their ambitious goals. On the other hand the company emphasizes the following three results as their most important gains:

- Change to cell production with:
  - visualization of stocks and demand
  - more efficient and flexible shop floor organization.
- Reduction of setup times in many cells (60-85% reduction).
- Change of attitude and involvement - people are now capable and willing to do continuous improvement.

Their measurements show stagnation regarding the chosen metrics during 1993 and an internal joke says that only the degree of KANBAN management is increasing. This lack of results is becoming a problem for the pioneers in the project. However, in the same period they introduced a new product line inside the existing buildings that would have been impossible without the change in layout. They reduced the delivery time and offered better delivery flexibility for their customers. Measurement and evaluation are obviously mixed and results are compared directly without taking into account that they already have distributed the gain by giving the customer better delivery flexibility. A stable delivery precision, stable value of tied capital and throughput time could then actually be an improvement in performance.

4.4 DEVELOPING A PERFORMANCE MANAGEMENT SYSTEM

Our introduction has revealed a need for change in the Western management approach. We will emphasize performance management as a new approach. It is perhaps not appropriate to claim it to be a new approach but rather an integration of already known techniques and methods. Everyone within an organization from top manager down to each single operator has two main tasks to fulfill (Sink and Tuttle, 1989):

- Get the job done right (that means done on time, within quality specifications, and the right amount of resources)
- Continuously improve individual, group, organizational, and system performance

This implies that everybody is supposed to continuously change their performance. A continuous change process has to be monitored to keep the process on the track toward the vision. Decentralized change combined with larger centralized improvement projects requires new management methods to ensure consistent development. Performance management is an active approach to guide the journey toward the vision.

Business performance has to be extended from financial efficiency to include customer satisfaction and future potential. An isolated financial result could lead to wrong conclusions about real performance, when the ultimate goal is long-term survival.

> A Russian director of a large agricultural collective received the productivity award for production of meat during subsequent three years. The fourth year he shot himself. His successor soon discovered that the collective had no breeders anymore.

Maintenance, R&D, training and development of the organizational culture are examples of costs that could be reduced with short-term improvement of profit. However, the long-term consequences could be devastating. Performance could be defined by three dimensions in a business context as shown in Figure 4.6 (Bredrup, 1994a):

- Effectiveness - to what extent are customers needs met;
- Efficiency - how economically are the resources of the company utilized;
- Changeability - to what extent are the company prepared for future changes.

The integration of these dimensions will ultimately decide the

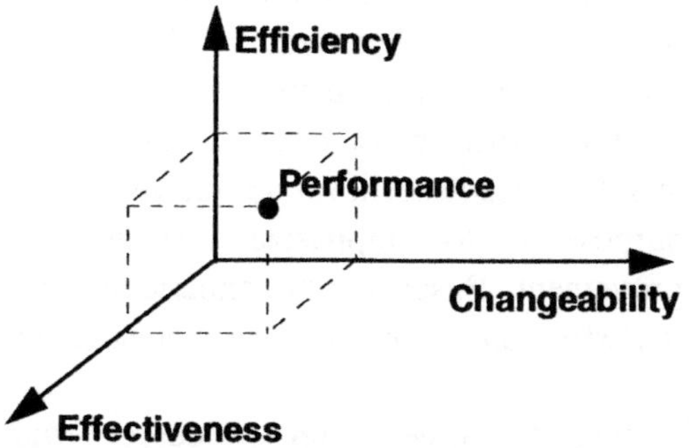

Figure 4.6 Dimensions deciding performance.

competitiveness of a company. Management approaches then have to be changed to support development of performance in all three dimensions.

The industrial environment has changed dramatically in the past decades and will change even more rapidly in the future. This requires a total re-thinking of how management is performed. Sink and Tuttle (1989) have defined performance management as a process containing:

1. Creating visions of what the desired future state is;
2. Planning - assessing present organizational status relative to the vision, creating strategies for how the desired future state can be attained, and building on strengths so as to move toward your visions;
3. Designing, developing, and effectively implementing specific improvement intervention that have a high probability of moving us toward the desired future state, particularly in terms of levels of performance;
4. Designing, redesigning, developing, and implementing measurement and evaluation systems that will tell us whether we are going where we said we wanted to go and how well we are doing along the way;
5. Ensuring that cultural support systems are in place so that we are rewarding and reinforcing progress. This is necessary to maintain the excellence we are achieving and to control levels of performance necessary to compete in future.

Many authors and practitioners claim that management is a profession and that for instance a military leader could easily run a manufacturing company. We strongly disagree with this. If management is a profession then it is reduced to a matter of information perception and decision making. A professional manager is then supposed to have unique abilities to catch information right and make the optimal decisions. This approach may be successful in a stable environment, but will probably fail in a rapidly changing environment. The reason why is that it assumes the existence of a well-defined infrastructure able to portray all necessary information. This infrastructure is often based on the organizational structure and is therefore obsolescent for the future.

Our main purpose is to emphasize managers' responsibility for performance improvement. Based on the experience discussed so far we have developed a performance improvement management model as shown in Figure 4.7.

This model suggests three main processes within performance management: performance planning, performance improvement and

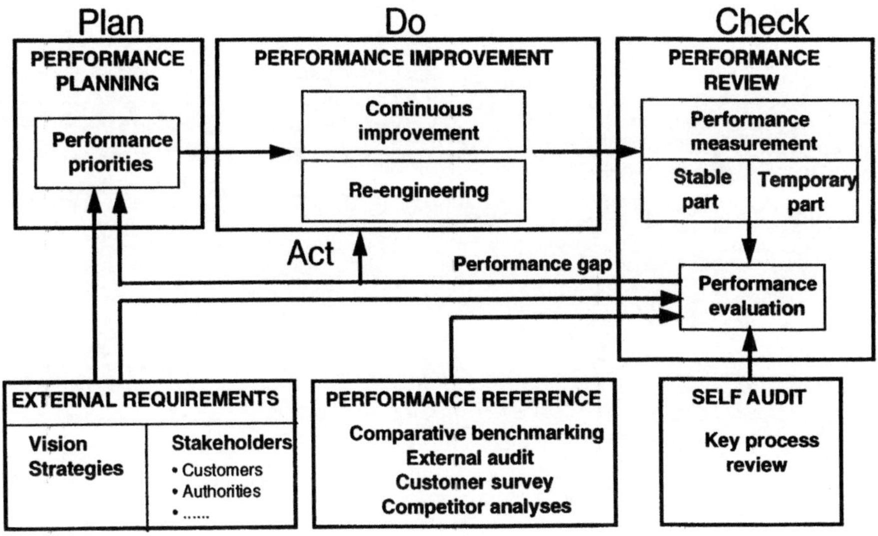

Figure 4.7 Performance management model

performance review. We use the term process deliberately because they are continuously ongoing processes and not pure sequential activities as illustrated by Figure 4.8. It should be stated that these processes are closely linked and do not implicate different responsibilities.

The performance management model, the main activities and the dynamic of the improvement cycle will be discussed in detail in the throughout this book.

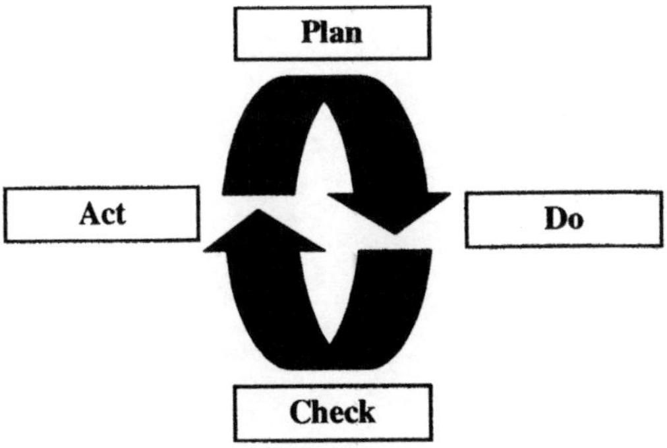

Figure 4.8 The plan-do-check-act circle.

PART TWO

Performance Planning

5. THE TRADITIONAL PLANNING HIERARCHY

In a situation with rapid change people may question the benefit of planning. In their study 'In search of excellence,' Peters and Waterman (1982) emphasized strategic planning integrated with a good understanding of own competitive advantages as the most important success factor. Jauch and Glueck (1988) states that businesses with a formal strategic planning process have a higher probability of success than those without a formal process. On the other hand there exists no such thing as a definite study that proves that strategic management causes better performance.

There exist three kinds of companies: those who makes things happen, those who are watching that things happen and those who wonder what happened. There should be no doubt that companies in the first category have the best chance to survive.

Perfect knowledge about future development is beyond reach in most areas. However, better knowledge of what the future will bring could help companies to prepare better for future action than competitors. These actions again help them influence their future. The time factor plays an extremely important role for the preparation of decisions and actions specially concerning competitive advantage. Many decisions require long-term preparation and success depends on timing. However, two basic questions requires a well-defined answer from every company:

- What business should the company be in?
- Why is the company in business?

The planning hierarchy should answer these questions properly. It could be defined in several ways depending on purpose and basis. Sink and Tuttle (1989) defines the planning hierarchy as follows:

- Vision
- Mission (purpose)
- Guiding principles (values and beliefs)
- Superordinate goals

- Goals
- Objectives
- Activities

5.1 VISION

A vision indicates what the company is aiming at in the future. It serves as the long-term road map for the company. Two statements of old wisdom could illustrate the role of visions:

> *If you do not know where you are going, any road will take you there.*
>
> *If you do not know where you are, a map is unlikely to help you.*

Creating a vision means that the company wishes to influence its future. The time horizon of the vision could be up to 20 years. Many will argue against such a long time horizon for planning due to the increasingly rapid change cycles most industries are facing. Changes will make the vision obsolete, but exact fulfillment is not the point with creating a vision. The main purpose is to develop guidelines for all planning to ensure consistency to purpose. Scenarios for what decides future success will change over time and the vision needs to be updated.

The time horizon for the vision could vary from industry to industry. Product development is an example of the need for differences in time horizon. Product development projects for airplanes have a time horizon of 12-15 years whereas computer companies could have several market releases every year. An electronics company obviously experiences rapid change cycles in both product and production technology and subsequently could operate with a shorter time horizon for commitment. However, product life cycle is not enough as a criterion. Consider an electronic company implementing the Total Quality Management philosophy. Successful implementation requires a devoted organization and a long-term commitment. Success is hard to achieve even if all members in the organization are supportive. An absolute requirement is that no conflicting strategies and actions are implemented during the process. The product range has probably been totally changed during this process. This means that the planning horizon has to exceed the most long-termed process to ensure constancy to purpose. Examples of long-term processes could be implementing JIT, changing production technology from line production to flexible manufacturing and a process towards a lean supply chain.

A supplier for the automotive industry wanted to develop system solutions for their customers instead of just being a manufacturer of customer specified parts. The time horizon before mass production is at least 12 years. This vision has serious consequences for manufacturing technology and the company had to gradually change from traditional line production to more flexible manufacturing equipment. All intermediate plans have to be in accordance to the vision.

The vision is a description of the organization of the future. It will describe the most important characteristics that determine whether the company is in business or not. From a macro productivity point of view it is important that every effort brings the company in the same and preferentially right direction. A key question for the vision is to define the 'right' direction.

Definition of a vision depends on several factors as for instance:

- Customer needs
- Management ambitions
- Competitors' initiatives
- Technological development
- Financial needs
- Political decisions
- Trade regulations
- Changing legislation
- Environmental constraints

Major changes in these factors require a change in the capability to stay competitive. Ability to forecast future changes have always been looked upon as a valuable quality. Unfortunately, forecasting is too difficult for most people. A natural consequence of accepting changes as inevitable is to improve the ability to change. However, the ability to adapt to changing environment depends on a various number of factors like organizational culture and characteristics of facilities and machinery. A vision could serve as a guiding star for the development of strategies and operational plans to improve the ability to change. No single strategy will be able to finish this process and a vision is required to ensure constancy to purpose.

Figure 5.1 Miraculous fulfillment of the vision.

Introduction of broad-based improvement projects in many companies is characterized by an unrealistic opinion of how to fulfill their ambitious visions as illustrated in Figure 5.1.

Most companies realize that the future will require improved performance, but fail to respond properly. They see new emerging philosophies like Total Quality Management, Total Productive Maintenance, Just-in-Time, Time-Based Management and Activity-Based Management as a quick fix. They simply fail to acknowledge that successful implementation of these philosophies requires a long-term horizon. Results are often not available in removal of existing costs, but as an improved ability to change and hence a reduction of future costs. Without an accepted vision the forces against change could use poor results as ammunition to cut off the projects.

5.2 MISSION

A mission defines the scope of the business activities the company pursues. Kotler (1988) states that the mission statement should cover:

- Market segment scope
- Industry scope
- Technology scope
- Vertical scope
- Geographical scope

Jauch and Glueck (1988) emphasize the importance of legitimacy. Long-term survival depends on ability to legitimatize the company in the market. However, defining a meaningful mission is a very difficult task because there are plenty of pitfalls. A mission statement could be so broad that it becomes meaningless or even so pompous that it becomes ridiculous. Defining a powerful mission depends on finding the right balance between visions and specific guidelines.

Jauch and Glueck (1988) states that the mission has to be clear enough so that it leads to actions. Some kind of targets has to be established so that progress could be evaluated. This does not necessarily implicate use of detailed measures, but is meant as a guideline to ensure commitment to the mission. A mission should serve as guidelines for strategic decisions.

A mission answers the question: 'What business should the company be in?' It could be argued that the mission actually is the top level of the planning hierarchy. This is perhaps the difference between a pure marketing view and our performance view of the planning hierarchy. From our point of view, defining the scope of the business depends on the capabilities the company possesses. Definition of market scope has to be in accordance with long-term capability of the company. As life cycle for products decreases it becomes more important to develop products that fit the capability to ensure profitability. We do not argue for a renaissance of the technology-driven company, but advocate the need to exploit the capability of the company in the best possible way. Major changes in the structure of a company are not possible in the short run to satisfy short-termed market demands. However, the company can improve their ability to respond on changes in the environment as a step towards the vision.

5.3 GUIDING PRINCIPLES

Cultural aspects are of major importance for all planning, because successful implementation often requires a change in the organizational culture. Culture could be defined as:

> *The pattern of basic assumptions - invented, discovered, or developed by a given group as it learns to cope with its problems of external adaptation and internal integration - that has worked well enough to be considered valid and, therefore, to be taught to new members as the correct way to perceive, think and feel in relation to those problems.*

A critical task for performance management and hence performance planning is to create and shape a supportive organizational culture. Every company has an organizational culture, but few are aware of what constitutes their culture and how it could be changed. It is not sufficient to develop new policies. The informal part of guiding principles is often dominating the culture. Implementation of philosophies like Total Quality Management and Just-in-Time are impossible if the culture does not adapt to new requirements. Deal and Kennedy (1985) have identified five key elements of culture:

- **Business environment**. Every company is influenced by the reality in the market, depending on the products, competitors, customers, technologies, authorities, etc.
- **Values**. They are the basic views and beliefs in the company. Values define success in specific terms for the employees.
- **Heroes**. The people that personify the values and hence are ideals for the employees.
- **Rituals**. The systematic routines in the company
- **The cultural network**. This network is the carrier of the values. A cultural network has several role characters like storytellers, spies, 'priests', cliques and whisperers. To work efficiently within the network is the only way to get things done fast or to understand what is really happening.

Deal and Kennedy (1985) have identified four main groups of cultures. They are based on the degree of risk connected to the business activities and how fast feedback on success of decisions or strategies is available:

- **The tough and dynamic culture**. It is characterized by a world with individualization, high risk and rapid feedback. Examples of industries are consulting companies, entertainment industries and construction and building.
- **Work hard/play hard culture**. Fun and action are the rule and the employees take few chances, all with rapid feedback. The culture encourages people to maintain a high level of activity to succeed. This is often companies with sales to a big public where the sales activity is the most critical business process.

- **Invest in the company culture**. This is a culture with very high risk and very slow feedback. This culture is common in manufacturers of capital goods like aerospace, cars and industrial machinery.
- **Process culture**. A world with little or no feedback where employees have problems with measuring what they are doing. Instead they concentrate on how things are done. If the process is not managed it could be called bureaucracy. Examples are banks and other financial companies, very regulated industries and most public companies.

Identifying the existing culture is a very profitable activity, because it reveals the invisible but nevertheless destructive conflicts between individual goals and goals for the company. Removing these conflicts and aiming at correlation between individual and company goals are prioritized activities within performance management. Most strategies will fail if a simultaneously change of culture is not achieved.

The business environment is the most important factor determining the culture. A critical task is to visualize the connection between the requirements from the environment and the culture to promote change. However, there are other sources for a mismatch between the culture and the desired capability like for instance measurement and incentive systems. These sources have to be detected and actions to develop new guidelines within the company have to be planned.

5.4 STRATEGIC PLANNING

Strategy is originally a military expression and the original meaning is, according to *The Random House dictionary of the English language*:

> *The science or art of combining and employing the means of war in planning and directing large military movements and operations.*

In military usage, a distinction is made between strategy and tactics:

> *Strategy is the utilization, during both peace and war of all of a nation's forces, through large-scale, long-range planning and development, to ensure security and victory.*

> *Tactics deals with the use and deployment of troops in actual combat*

In a business context a strategy will include decisions like:

- What products on what markets
- What technology
- Which resources

Kotler (1988) defines strategic planning as follows:

> *Strategic planning is the managerial process of developing and maintaining a viable fit between the organization's objectives and resources, and it's changing market opportunities. The aim of strategic planning is to shape and reshape the company's businesses and products so that they combine to produce satisfactory profits and growth.*

Correctly he defines this as the art of market-oriented strategic planning. Recent literature with other background emphasizes the lack of focus on operational aspects. The 1992 *US Manufacturing Futures Survey* (Kim and Miller, 1992) showed significant changes in the strategic role of different functions as shown in Figure 5.2. The 1987 and 1992 scales measure how significant a function's role is/was in setting business strategy with 7 as very strong and 1 as insignificant. The future scale is whether a function's strategy will increase or decrease in the future. This survey indicates a change in relative importance from marketing/sales towards manufacturing and engineering/R&D in the future.

Competitive advantage is gained through products or processes. Innovations in product and production technology and productivity improvements are important for sustaining competitive advantage. Hayes and Wheelwright (1984) are very concerned with the lack of production strategy and emphasize this as a main reason for the American loss of competitiveness. Jauch & Glueck (1988) define strategy as follows:

> *A strategy is a unified, comprehensive and integrated plan that relates the strategic advantages of the firm to the challenges of the environment. It is designed to ensure that the basic objectives of the enterprise are achieved through proper execution by the organization.*

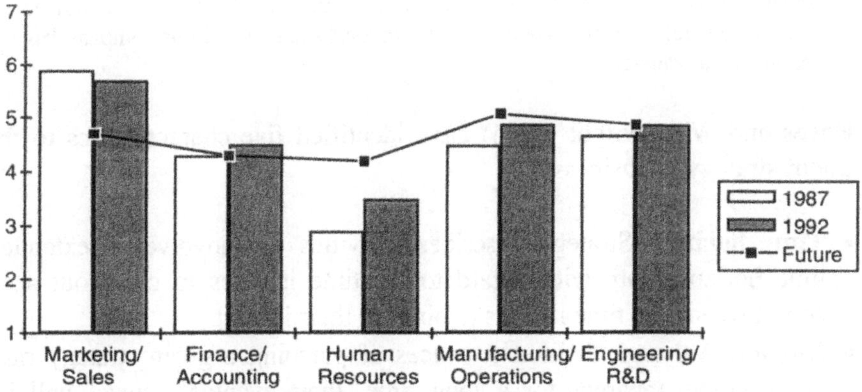

Figure 5.2 Who influences strategy development?

With unified it means that it ties all the parts of an enterprise together. Comprehensive means that it covers all major aspects in an enterprise. To be integrated means that all the parts of the plan are compatible with each other and fit together well. This definition is very ambitious and clearly states that strategic planning involves more than marketing. Nobody questions the importance of marketing, but the strategic planning has to be balanced and long-term development for the company as a whole emphasized. Product/market requirements have to be matched/integrated with the process technology and long-term capacity strategies, facility strategy, integration strategy and so forth which have to be developed. Major changes require long lead times and long-term planning is essential.

Process technology or capacity is not changed over night, but long-term adjustment is necessary. Integration of all aspects in the strategy is necessary to ensure effective use of resources and avoid suboptimization. World-class manufacturers are sitting in the driver's seat and manage their evolution. Followers could at best hope to detect the need for change and adapt at an early stage. World-class manufacturers are always a step ahead when the followers adopt new methods or techniques. It's certainly better to act out of strength than out of weakness and world-class competition gives no honor for silver medalists.

A Norwegian subcontractor for the automobile industry said to their customer that acceptance testing was burglary. They offered to pay all related costs if failures were detected later in the process. The subcontractor was so confident about their own level of quality that they wanted to exploit it as a competitive advantage. This move was a success. The customer adopted it as a standard and it most certainly lead to a better relationship and confidence between customer and subcontractor. Competitors are not ready for this, but could be forced to offer the same. Their costs are probably higher and our company has gained an advantage.

Hayes and Wheelwright (1984) have identified five characteristics to the use of strategy in business:

- **Time horizon**. Strategy describes activities that involve an extended time horizon, both with regard to the time it takes to carry out such activities and the time it takes to observe their impact.
- **Impact**. Although the consequences of pursuing a given strategy may not become apparent for a long time, their eventual impact will be significant.
- **Concentration of effort**. An effective strategy usually requires concentrating one's activity, effort, or attention on a fairly narrow range of pursuits.
- **Pattern of decisions**. Most strategies require that a series of certain types of decisions be made over a period of time. These decisions must be supportive of one another, in that they follow a consistent pattern.
- **Pervasiveness**. A strategy embraces a wide spectrum of activities ranging from resource allocation processes to day-to-day operations. In addition the need for consistency over time in these activities requires that all levels of an organization act in ways that reinforce the strategy.

Strategic planning has traditionally been a task for top management. There has been little willingness to involve others in the process. Confidentiality has been emphasized due to the belief that competitors could catch the strategy and benefit from the knowledge. However, confidentiality often confuses middle managers and other employees more than the enemies. Strategic planning as practised by most organizations is characterised by (Sink and Tuttle, 1989):

- formal and bureaucratic;
- focused on the plan and ignoring the process;
- is budget driven (the budget drives the plans);
- involves top management and their consultants;
- is overly influenced by marketing, business acquisitions and finance considerations;

- is myopic and short-sighted due to limited involvement and unrealistic expectations for quick and easy gains;
- is often detached from many pragmatic operational realities.

The need for change in strategic planning could be summarized as:

- involve more people;
- be structured but less formal;
- focus on the plan and the process;
- drive the budget;
- develop better balance between marketing, finance and operations;
- lengthen the planning horizons;
- more effectively link action planning and effective implementation.

The planning process needs to involve considerably more people than today. Effective implementation requires feelings of ownership and involvement by those who are finally going to implement the plan. Those involved in implementation must also play a part in the development process. People in the organization have ideas and experience that could be valuable input to the process. The process is often more valuable than the plans as cited from Dwight D. Eisenhower:

> *Plans are nothing, planning is everything*

To facilitate involvement of more people and have an effective and efficient process at the same time, the process will have to be more structured but less formal than today. Simply sitting down around the table and hoping that a plan will eventually surface before the end of the meeting will not do. There will have to be guidelines to follow on how to run the process.

Furthermore, there is a need for loosening the tight connection between budget and plan. The traditional approach is to start with the budget, usually by extrapolating last year's budget. Then the plan is worked out within the boundaries set by the budget. The dominating question in such sessions is 'Can we afford this?' In the light of the rapid pace of change the question should be 'Can we afford not to do it?' Being allowed to develop the plan without constantly banging your head on the budget wall is crucial to the planning process. The budget is, and should be, a result of the plan, and not vice versa.

Finally, a shift has to occur regarding the areas focused on in the strategic plan. We have to move away from the heavy emphasis on marketing, business acquisitions, and financial considerations. Operational issues, and particularly the improvement of such, will have to be given far more attention in strategic planning.

There exist several methods of performing a strategic plan satisfying the new requirements for the process. Sink and Tuttle (1989) give an operational description of an eight-step model. We will focus on the integration of customer satisfaction in the strategic plan through performance planning.

6. PERFORMANCE PLANNING TO ENSURE BUSINESS ACHIEVEMENTS

We will introduce a planning model that enables companies to integrate business and performance strategies. The relation between strategies and business achievements are often difficult to describe. This model defines performance from an external point of view and provides a direct link from stakeholders' requirements to business requirements. In addition, adaptability to long-term visions and business strategies are integrated in this process and define performance requirements together with stakeholders' requirements.

The model is general and could be applied for any stakeholder of importance. However, we will focus on customers assuming that customer satisfaction is a major key to business achievements. Our model is based on the Quality Function Deployment methodology, which is originally developed as a customer-oriented approach to product innovation. In performance planning QFD is extended to cover both stakeholders and strategic requirements to business performance.

A stepwise implementation of this model is preferable and integration of customer requirements is first in the line for all companies. Customers are then the main focus in our presentation.

6.1 CUSTOMER SATISFACTION AS A PROFIT STRATEGY

Customer satisfaction is a focused topic in most disciplines regarding business performance and especially emphasized in marketing and quality management. Operationally, customer satisfaction could be defined as shown in Figure 6.1.

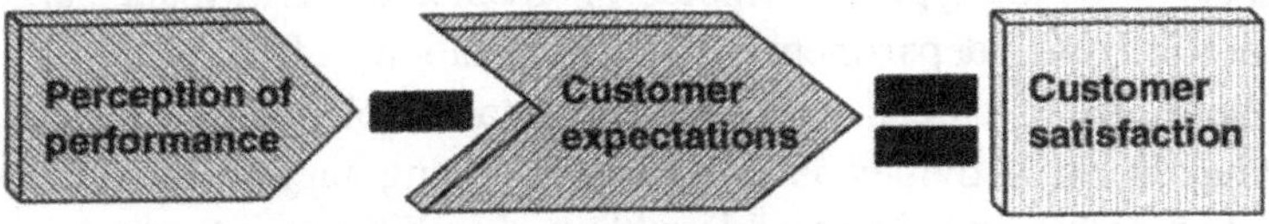

Figure 6.1 Definition of customer satisfaction.

Since customer satisfaction is the difference between performance and expectations, efficient management is only possible through good knowledge of customer preferences and experience. Both performance and expectations are characterized by evaluation. Exact knowledge is unachievable. Preferences and expectations are dynamic and influenced by a number of factors:

- Advertising from the company or a competitor
- Recommendations from others
- Promises from a salesperson
- Published specifications
- Industry norms
- Standards
- Previous experience with the product or service
- Experience with a competitor

Creating and managing customer expectations is an important aspect of marketing whether the case is product introduction or market penetration. When a company gets an order, it means that the customer has higher expectations to the company than to competitors. Customer satisfaction depends on the ability to fulfill or even exceed these expectations. However, creating unrealistic expectations is dangerous, because dissatisfaction may lead to change of supplier next time.

> A manufacturer of office chairs promises maximum five days delivery time for standard chairs all around Europe. Competitors are not able to promise less than 15 days. The company has recently experienced problems with keeping the promised delivery time and several customers have experienced three days delay. Despite the fact that this is still far below competitors' delivery time they have received several complaints and the customer satisfaction level has decreased.

This example clearly shows that broken promises may have devastating consequences. Delivery precision could be more important than delivery time for some customers, leading to a loss of these customers.

Expectations are typically created in advance of performance and often involve different departments. Synchronization of expectations and performance is a crucial task in performance planning. Integration of marketing in all activities is a necessity. Using targets in a customer-oriented organization is impossible without the ability to integrate.

According to Giffi *et al.* (1990) focus on customer satisfaction is one of eight categories of 'best practices' that identify a world-class manufacturer. A satisfied customer (Kotler, 1988):

- Buys again;
- Talks favorably to others about the company;
- Pays less attention to competing brands and advertising;
- Buys other products that the company can later add to its line.

However, most companies are concerned with measurable numbers and often fail to estimate the significance of the more intangible indicators. Deming (1986) states that the most important numbers are difficult to measure and simply managing by the easy available numbers is a deadly disease. Kotler (1988) says something about the magnitude of the problem in the following; 'whereas a satisfied customer tells three people about a good product experience, a dissatisfied customer gripes to eleven people'. Kotler also refers to an American study done for the Carter administration in 1985 revealing that 96 percent of unhappy customer never tell the company. This shows that received complaints are probably only the tip of the iceberg. It's not enough to passively receive customer reactions to be customer-oriented.

Re-buy intention is an important key figure for most companies. This is especially important for industrial marketers. The number of new potential customers is considerable less and the cost of marketing per customer is much higher than in the consumer market. Re-buy intention is a major indicator for customer satisfaction and influences directly both turnover and margin. It costs several times more to win a new customer than to get re-buy from the existing customers. Customer loyalty is only achieved by satisfying your customer. Strategies to ensure customer satisfaction are therefore true profit strategies as illustrated in Figure 6.2.

Productivity is a duplex matter. It's about doing the right things with efficiency as high as possible. Customer satisfaction depends on customers' perception of performance. No matter how well the company fulfills their internally defined performance measures, perception of customers is all that counts. Doing the right things means ensuring customer satisfaction.

Manufacturing companies start to realize that to remain competitive, they must adopt practices from service companies. Some of the world-class manufacturers like Hewlett Packard, Caterpillar, and Allen-Bradley run their production groups like service factories (Lynch and Cross, 1991).

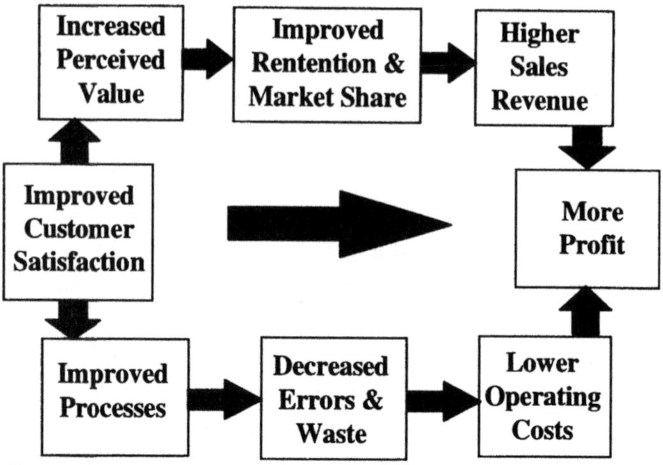

Figure 6.2 Customer satisfaction as a profit strategy.

Viewing products as services implies that business does not end by selling the product. A service begins with an order and ends only when the customer is satisfied. Adapting this perspective, makes it possible to identify critical points of customer contact with your business.

Each time a customer is in contact with a product, a service or a company, it is an opportunity to make a favorable impression (or to create dissatisfaction). A wide definition of contact also involves recommendations or warnings from other customers. A bad image will spread faster than a good one, and regaining an image is very time and cost consuming. In relation to this discussion, it could be worthwhile mentioning the definition of quality in ISO 8402:

> *Quality is the totality of features and characteristics of a product or service that bear on its ability to satisfy stated or implied needs.*

This definition clearly states that quality also includes satisfaction of implied needs. Identification of these implied needs are then a critical task of marketing. A customer-oriented company cannot allow distribution of attitudes like 'Why do customers always have inadequate access to their premises.' The manufacturer should be aware that they are not just selling a physical product, but selling a solution of a need or a problem. Kotler (1988) defines a product as follows:

> *A product is anything that can be offered to a market for attention, acquisition, use, or consumption that might satisfy a want or need.*

Most physical products contain a degree of service. Only the most standardized and cheapest commodities could be free of service. The degree of service is closely related to differentiation. Kotler (1988) divides a product in three levels as shown in Figure 6.3:

- *Core product.* Describes what the buyer is really buying and clearly states the benefit for the customer.
- *Tangible product.* Characteristics describing the tangible product: quality level, features, styling, and packaging.
- *Augmented product.* Factors like delivery terms, installation, after sale service and warranty could be part of the augmented product.

As competition increases, differentiation is probably only achievable in the peripheral parts of the product (we neglect unique product technology in this discussion). This is due to the dynamic of competition. When a company has gained competitive advantage with differentiation, cost leaders will try to equalize the advantage by standardizing the differentiation criteria. High quality levels according to the definition of conformance to specifications, is an example of a standardized differentiation criterion.

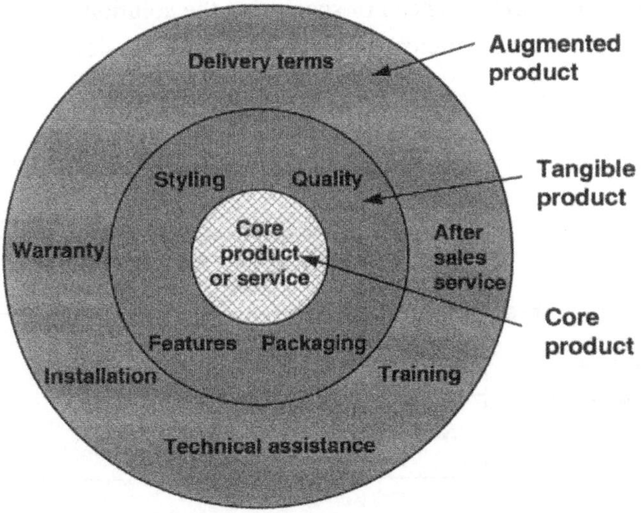

Figure 6.3 Definition of a total product.

The Kano model is a response to the fact that most products and primary services have become so common that special attractions are increasingly important for the customer quality evaluation of a company. It divides customer's response to quality attributes into three dimensions (Aune, 1993):

- Inevitable (implied, inevitable and forgotten attributes)
- Satisfiers (outspoken, measurable and technical attributes)
- Attractions (unrecognized, unspoken, trend-sensitive and customer-adjusted attributes

The level of these three decides the customer satisfaction. A high customer satisfaction is only possible with a high level of satisfiers and attractions. However, the absence of an inevitable attribute automatically creates dissatisfaction no matter how many attractions are offered. This relation is shown by O'Neal (1992) in Figure 6.4.

There are several sources for process improvement through adapting methods from service. Service production has important characteristics that could be useful for manufacturing industry:

- Services have a limited ability to buffer customer demand with inventory, therefore they must be very responsive;
- The customer is often involved in the actual service creation process;
- As products increase in complexity customers respond by specifying the actual service required instead of specifying the solution;
- The share of service is increasing in deliveries of physical products.

<table>
<tr><td colspan="2"></td><td colspan="3">Quality attribute perceived as fulfilled?</td></tr>
<tr><td colspan="2"></td><td>No</td><td colspan="2">Yes</td></tr>
<tr><td rowspan="3">Quality attributes expected?</td><td rowspan="2">Yes</td><td rowspan="2">Dissatisfied</td><td>Expectations met</td><td>Expectations exceeded</td></tr>
<tr><td>Satisfied</td><td>Delighted</td></tr>
<tr><td>No</td><td>No effect</td><td colspan="2">Delighted</td></tr>
</table>

Figure 6.4 Model of expectation/fulfillment outcome (O'Neal, 1992).

When an automotive company specifies the need for a paint robot, they just say that they are buying a painted door and not the technical specifications of that robot. The specifications simply state the required characteristics of the service like thickness of the paint and setup times. Then it is up to the bidders to offer the best solution.

Each contact with a customer could be viewed as a service and is an opportunity to either satisfy or dissatisfy the customer. Contacts are split in two categories:

- Primary points of contact are directly related to the consumption of the product or service;
- Secondary points of contact are those not directly related to the consumption itself.

For each customer contact two basic questions could help identifying the attributes of performance that will satisfy (or dissatisfy) the customer:

- What constitutes good performance?
- What constitutes bad performance?

Viewing the business from the customers' perspective makes it easier to understand what needs to be done or to be avoided to maximize benefit for your customer. Identifying the critical success factors is not enough. The challenge is to manage the activities necessary to support the customer.

The *International Quality Study* (Harrington, 1993) indicates an increasingly importance of customer satisfaction in strategic planning as shown in Table 6.2. The numbers correspond to the percentage of businesses in the IQS study. Customer satisfaction is most focused in Japan both in the past, today and in the forseeable future. However, the importance as a primary criterion is increasing dramatically in the future for all countries in the study.

Table 6.2 Integration of customer satisfaction in the strategic planning process (IQS).

Country	Past	Present	Future
Germany	10%	22%	59%
Japan	30%	42%	80%
US	18%	37%	69%

6.2 DETERMINING THE CHARACTERISTICS OF PERFORMANCE

Performance has been previously defined by three dimensions: efficiency, effectiveness and adaptability. We will provide a methodology to define performance from an external point of view as shown in Figure 6.5 (Bredrup and Bredrup, 1994a). By defining performance based on the stakeholder model and the vision and strategy we achieve a direct link to competitiveness. However, it is important to know the existing priorities and indicators to develop a common understanding of priorities. Priorities are a part of the organizational culture and implementation based on a top-down approach without involvement are likely to fail. We will emphasize the need for a combination between the top-down and bottom-up approach.

Unless this process is forced by a serious crisis, it is almost impossible to design completely new priorities with a supporting measurement system at once. Our methodology could be applied in a total re-engineering of the management model, but we will strongly emphasize the use of pilot cases to make involvement possible. Achieving a common understanding of priorities is the most critical task for future efficiency.

This model integrates existing external requirements and future requirements. Efficiency and effectiveness are covering existing requirements whereas future requirements are covered by adaptability. Efficiency is the result of the match between external requirements and the

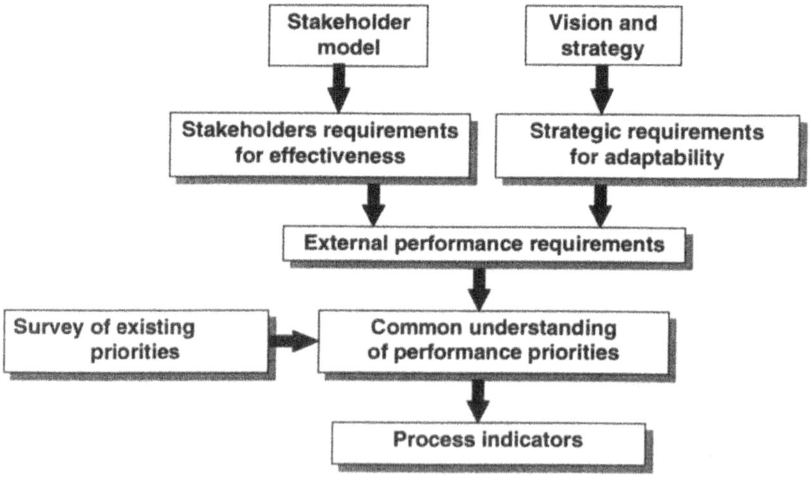

Figure 6.5 Model for definition of performance.

existing structure. Looking at performance requirements from an external point of view is required to balance all three dimensions of performance.

6.2.1 Performance requirements from stakeholders point of view

Long-term survival of a company depends on satisfying stakeholders needs. In the previous chapter we have emphasized the importance of customer-oriented performance requirements but stakeholders also include employees, suppliers, shareholders, alliance partners, parent company, environment, etc. To ensure an adequate performance level perceived from each stakeholder, performance requirements have to be defined from the stakeholders point of view. Competitiveness depends on attractiveness towards the different stakeholders and performance has to be planned to increase this attractiveness.

Stakeholders have interest in special aspects of performance and conflicting interests especially regarding distribution of productivity improvement may exist. Identification of single indicators to satisfy the different stakeholders may be counterproductive and lead to suboptimization. There exists no ideal solution on such trade-off problems, but some approaches may be better than others. A systematic approach to identify stakeholders' requirements may reveal opportunities for performance. Stakeholder satisfaction is determined by the difference between expectations and perceived performance. Perceived performance is then related to the interfaces between the company and its stakeholders. Definition of performance requirements from a stakeholder perspective could then follow the sequence as seen in Figure 6.6 (Bredrup and Bredrup, 1994a).

1. Define the stakeholder model with relative importance
2. Identify interfaces between stakeholders and company
3. Identify performance requirements from the stakeholders' perspective
4. Define importance of each requirement

Define the stakeholder model with relative importance

A company has relationships with a number of stakeholders. The list of stakeholders includes groups like customers, suppliers, authorities, owners, financial institutions, employees, management, environment and alliance partners. Within each group there may exist several subsets with different characteristics. Application of the stakeholder model means accepting the

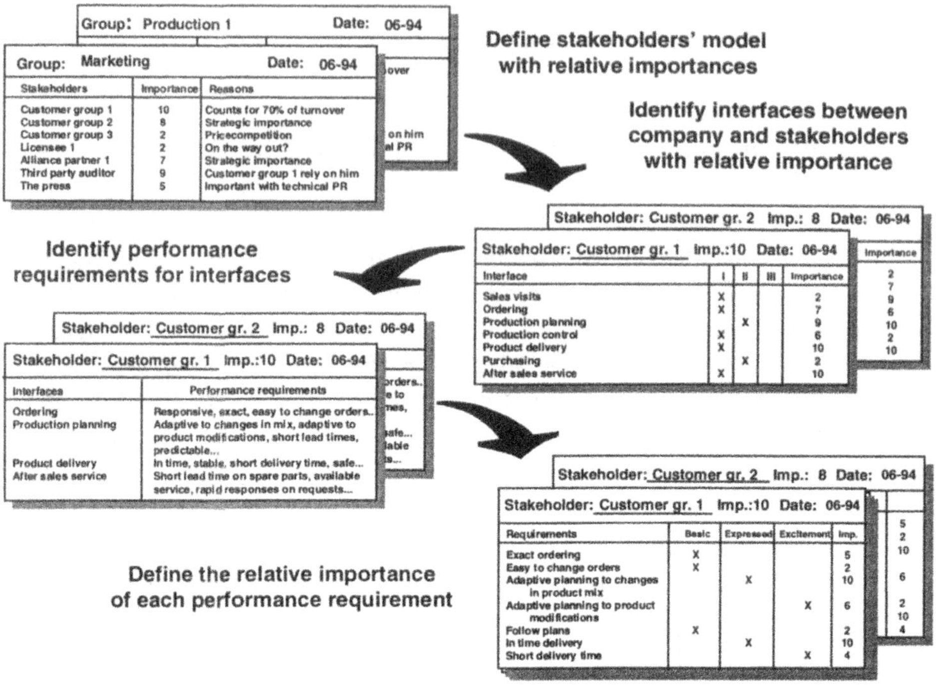

Figure 6.6 Process for defining stakeholders' requirements.

fact that a company and its stakeholders have mutual interest and dependency concerning specific business processes.

Identification of the set of stakeholders is the first step. Most stakeholders are easily identified due to direct business relations whereas more indirect stakeholders could easily be forgotten. Brainstorming is a useful tool to provide a representative set of stakeholders. All stakeholders with influence on competitiveness should be found. This is a creative process, and no sorting should occur. In some cases it could be wise to involve groups from different departments to ensure that all important stakeholders are identified. Charts like those shown in Figure 6.7 are useful to carry out an internal survey.

The different stakeholders have different impact on competitiveness. Assessment of importance is necessary to achieve a balanced understanding of the priorities for the business. Figure 6.8 is a reminder of the necessity in order to maximize cost/benefit of investments in performance improvement. This model shows that it is waste to provide high performance of characteristics of low importance. The most efficient strategy is to improve performance of characteristics of high importance to competitiveness.

Group: Marketing 1 Date: 06-94

Stakeholders	Importance	Reasons
Customer group 1	10	Counts for 70% of turnover
Customer group 2	8	Strategic importance
Customer group 3	2	Pricecompetition
Licensee 1	2	On the way out?
Alliance partner 1	7	Strategic importance
Third party auditor	9	Customer group 1 rely on him
The press	5	Important with technical PR

Figure 6.7 Example of a chart for identification of stakeholders.

Assessment of importance is a difficult task in complex companies. However, to desist from giving importance, is to give equal priority to everything. Objectiveness is impossible, but a well-prepared process of subjective evaluation is sufficient for our purpose. A process of evaluating importance of each stakeholder includes answering questions like:

- Financial impact of the stakeholder
- Availability of alternative stakeholders
- Attractiveness of the stakeholder
- Competitive position towards the stakeholder
- Investment in the relationship with the stakeholder
- Strategic importance

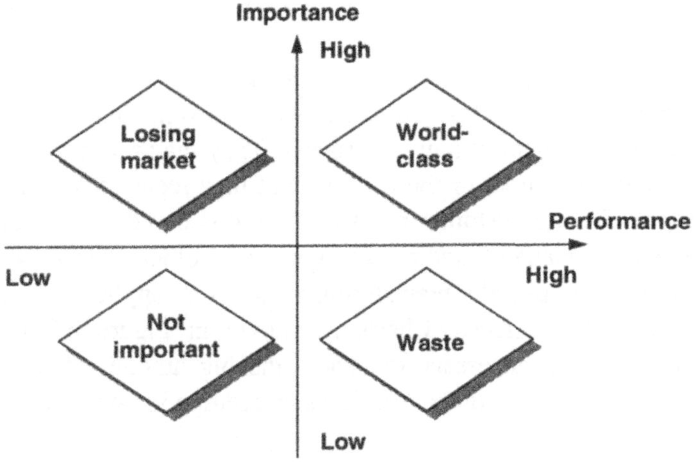

Figure 6.8 Relation between importance and performance.

It should be emphasized that this evaluation of importance is individual for each company and is a necessary step to understanding performance. Transformation of performance improvements into business achievements depends on the ability to identify the best prospects. Stakeholders of minor importance are not taken into consideration in the preceding process.

Identify performance interfaces between stakeholders and company

A stakeholder may have several interfaces to a company. These interfaces could have different characteristics and be of different importance. Identification of interfaces from the stakeholders' points of view will serve as a checklist for development of performance requirements. Three different kinds of interfaces may exist:

- Stakeholder and company are partners
- Stakeholder has direct interest in the output or the process
- Stakeholder has indirect interest in the output or the process

The first type is characterized by an active partnership in a process. A strong and lasting relationship between stakeholder and company is required. Intersections between the value chains of supplier/company and company/customer could serve as examples. World-class companies view their supply chain as an extension of their own value chain.

The second type includes interfaces were stakeholders have direct interest in the output of a process. A stakeholder may want to influence the output or the process itself despite not being an actor in the process. Examples may be customers' interests in quality assurance, environmental policy, production plans, documentation, etc. The main reason for this interest is to maximize benefit for the stakeholder .

The third type includes interfaces were stakeholders indirectly are concerned by the output of a given process. They may not have predefined opinions about the output or the process, but may respond negatively on a given output. From a customer point of view this could include personnel development, maintenance and marketing towards other customers.

Creative techniques like brainstorming should be applied to develop an extensive list of interfaces. Charts as shown in Figure 6.9 provide a systematic approach necessary to handle the big amount of information. The set is reduced by rating the interfaces by relative importance.

Stakeholder: Main customer — Importance: 10 — Date: 06-94

Interface	I	II	III	Importance
Sales visits	X			2
Ordering	X			7
Production planning		X		9
Production control	X			6
Product delivery	X			10
Purchasing		X		2
After sales service	X			10

Figure 6.9 Example of an interface chart.

Identify performance requirements

Performance has to be defined from the stakeholders point of view since their perception of performance determines the competitiveness of the company. Some stakeholders may have well-known requirements to performance as for instance finance institutions demanding a given interest rate for capital. Others may have dynamic or more vague expectations of the performance level.

Negotiations between company and stakeholders often regulate the financial aspects of the relationship, whereas more operational performance requirements are neglected. Access to continued education and training are examples of factors for employees' satisfaction that are often neglected in negotiations with unions. A stakeholder will typically have some outspoken and some implied requirements to performance. Figure 6.10 shows a model of the different categories of match between the stakeholders' and the company's perception of performance.

The company is responsible for collecting the information from the stakeholders. Traditional surveys are often not intended to collect this kind

		Company: Included in performance	Company: Not included in performance
Stakeholder	Outspoken	Obvious	Ignored
	Implied	Identified	Not identified
	Not intended	Waste or surprise	Unknown

Figure 6.10 Match between stakeholders' intention and performance.

of information, but could easily be adjusted to suit our purpose. However, most knowledge is already available within the company. A common problem is though that this information is distributed and related to individuals within the organization. A systematic approach is required to benefit from the knowledge of each individual.

The list of interfaces works as a checklist for development of stakeholders' performance requirements. Those involved in each interface have to systematize their knowledge. Additional information could be collected with surveys or other inquiries. Other sources for information both internally and externally may be tried. Customer surveys are further discussed in 6.2.2. Figure 6.11 is an example of a registration card.

Define importance of each requirement

At this stage we are reaching the essence of performance. By assessing importance of each requirement, a rating of performance priorities from the stakeholders' point of view is obtained. This importance rating is the key input to the performance planning model and has to be performed thoroughly. Information about these priorities is already available within the organization, but additional information has to be collected. Surveys or more informal questions could provide the necessary supplement.

The Kano model (Akao, 1990) in Figure 6.12 relates satisfaction to the degree to which performance requirements are achieved. This could be helpful in the assessment process. The straight line represents the expressed performance requirements from the stakeholder. In general, only the expressed requirements would be explicitly mentioned if the stakeholder is asked for input. However, there exists a set of basic requirements that are so obvious that they are not mentioned by the stakeholder. Together these two groups define the expectations of the stakeholder. The degree of perceived achievement decides the satisfaction level.

Stakeholder: Main customer　　Importance: 10　　Date: 06-94

Interfaces	Performance requirements
Ordering	Responsive, exact, easy to change orders..
Production planning	Adaptive to changes in mix, adaptive to product modifications, short lead times, predictable...
Product delivery	In time, stable, short delivery time, safe...
After sales service	Short lead time on spare parts, available service, rapid responses on requests...

Figure 6.11 An example of a registration card.

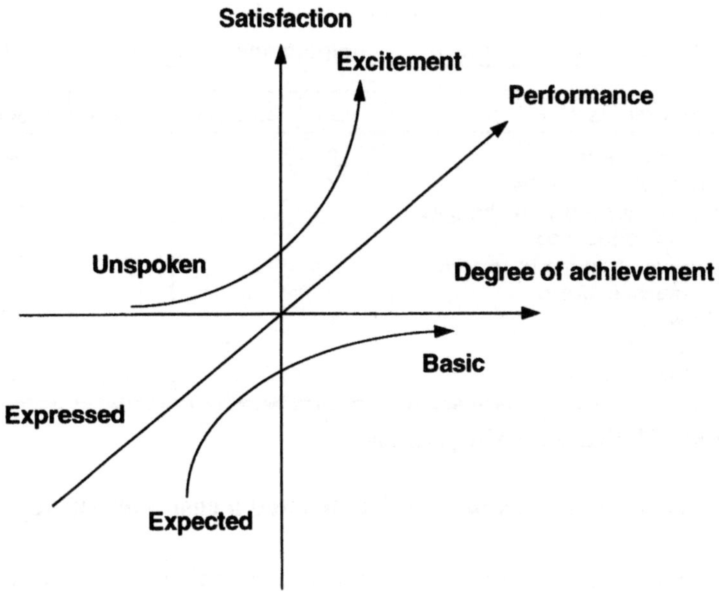

Figure 6.12 The Kano model.

Identification and fulfillment of expectations are necessary to achieve satisfaction. Satisfaction could be further enhanced by offering unspoken features. However, enhancement depends on achievement of basic expectations. Extra service will not equalize a scratch in the paint of a car.

The clear message is that identification and satisfaction of expectations is required before thinking of providing extra benefit. This is a guiding star for the importance evaluation process. Even if all of the basic requirements are met perfectly, we would not achieve real satisfaction. We would only eliminate dissatisfaction. It could be useful to categorize according to this tripartition of requirements as seen in Figure 6.13 .

Performance planning is a continuous process and not a sequence of individual events. Carrying out a new step in the planning process could provide new knowledge about previous steps, that should be included. New knowledge from other sources should also be included. The systematic approach makes it easy to update previous steps in the model.

Importance set for each requirement should not be seen upon as fixed. The relative importance between the different requirements is a matter of tactic. Tactics have to be frequently updated for instance if a competitor changes his tactics. Application of this performance planning model will help a company to make better choices.

Stakeholder: Main customer		Importance: 10	Date: 06-94	
Requirements	**Basic**	**Expressed**	**Exciter**	**Import.**
Exact ordering	X			5
Easy to change orders	X			2
Adaptive planning to changes in product mix		X		10
Adaptive planning to product modifications			X	6
Follow plans	X			2
In time delivery		X		10
Short delivery time			X	4

Figure 6.13 Evaluation of requirements.

6.2.2 Recording the customer voice through a customer survey

Surveys could be a useful tool to collect external information. We will discuss customer surveys in particular due to our devotion to customer-orientation. Our discussion is universal and could easily be applied to other stakeholders. In fact, surveying suppliers could provide useful input about benchmarking due to their experience with other competitors.

Most companies says they are recording customers' voice. However, most companies have a passive approach. At best they are recording what they are hearing or even worse, relying on the complaint rate. In both cases the company assumes that customers without comments and complaints are satisfied. Research has shown that this is totally wrong (Kotler, 1988). Only the extreme opinions are caught by a passive approach.

Objective measurement of customer perception is impossible, but precision is not essential. The objectives of a customer survey could be:

- Benchmark against competition
- Establish importance ranking
- Determine satisfaction level
- Determine re-purchase intent
- Analyze strengths and weaknesses
- Propose action plans

The ultimate goal of a customer survey is to identify sources for improvement of competitiveness. Ranking of importance is essential to identify areas for improvement. Improvement of areas of low importance could be a waste of resources. Several important decisions have to be taken when designing a customer survey:

- Time horizon
- Method
- Purpose
- Target group
- Organization
- Topics

The survey could be arranged as a continuous survey or as a study with a limited time horizon. This decision depends on several matters like:

- Purpose
- Number of customers
- Change cycles within the market
- Cost of the chosen survey method

There are several methods available for customer survey as:

- Interviews;
- Mailed questionnaire;
- Questionnaire answered when representatives visit the customer. (Carrying out the survey could be secondary.)

There is a clear difference between consumers and organizational buyers. The latter may have different people involved that complicates the link between performance and business success. However, one decision is common for both: what customers should be surveyed. Customers could be divided into a wide range of segments as for instance:

- Most profitable customers versus less profitable customers;
- Regular versus causal customers;
- Lost customers;
- Potential customers.

This decision is closely connected to definition of purpose. Surveying the most profitable customers is not enough because improvement is the main objective. Lost customers are an interesting segment because they have experience with competitors as well. Identification of reason for change is important in order to be able to improve. Improvement is necessary to avoid loosing more customers and perhaps win lost customers back. It could also be interesting to find out why some customers are less profitable than others.

A customer survey could be carried out by professionals or by people inside your own organization. Some think that customers are unwilling to answer such questions and hire professionals to avoid the bother. However, experienced people say that most customers are pleased by getting an opportunity to influence development. It is often an advantage to portray that the company is eager to improve. A professional company may reach customers your people are unable to reach, like for instance potential new customers, but the learning effect should not be underestimated. Customer survey is not solely for the marketing staff, but could also be done by operators and other employees in quiet periods or as a deliberate educational effort. Womack *et al.* (1990) actually makes a point out of using operators as salesmen in bad periods in his lean production concept.

6.2.3 Strategic performance requirements

Performance planning has to be integrated with the traditional planning hierarchy. Development of a long-term vision, indicating what the company should look like in the future, will provide useful input to performance planning. Strategic plans concerning new markets, new products, new marketing strategies, new technologies, etc, mean new performance requirements. These requirements have to be integrated with those from the stakeholder model so that both short-term and long-term business achievements are ensured.

Application of the performance planning model implies that requirements for performance have to be outlined both in the vision and strategic plans. Time horizon decides how detailed specifications have to be. Performance requirements for the vision have to be roughly described, whereas consequences of change in marketing mix have to be specified in more detail.

It is not controversial to emphasize a need for integration of all kinds of strategies. Problems arise concerning the variety of characteristics of the different strategies and lack of integrative tools. Performance planning is an integrative methodology, but depends on the ability to develop compatible performance requirements.

Development of performance requirements should have been a major concern in the feasibility study of different strategies. This is often a controversial issue between the production department and the marketing department. A striking example of this was found in a manufacturer of large engines.

The production department of a manufacturer of large engines was concerned about the increasing number of sales which demanded engineering. This was very costly, because no streamlined system was available to handle this. Due to a change in market segments they experienced shorter delivery times and tougher demands for delivery precision. Due to poor management of engineering they lost valuable time. They thought that this trouble was caused by poor sales methods and that it eventually would pass by. However, an interview of sales managers revealed that offering more tailored solutions was a part of the marketing strategy.

By applying the stakeholder model we have developed a methodology to identify performance requirements of the existing stakeholders, including customers. However, strategic decisions often imply dramatic changes in opportunities and new requirements for performance. Identification of these is essential to achieve performance improvement. A simplified example is shown in Figure 6.14.

New opportunities could include investments in new technology, suppliers with new characteristics, new management systems. Utilization of new opportunities depends on a mutual adjustment process with the existing system. Research (Bredrup *et al.*, 1994) has shown that investments in modern manufacturing often fail to provide expected benefit, because existing production management systems remain unchanged. Expected improvements in flexibility and reduction of lead times are not

Decision: Market flexibility   Importance: 10   Date: 06-94

Requirements	Importance
Be responsive on new orders	8
Able to change order on short notice	2
Direct transmission of orders	10
Short delivery time	1
Customization	10
Precise delivery	10
Prices remain constant despite improved flexibility	4

Figure 6.14 A simplified example of strategic requirements.

achieved due to lack of support in existing systems.

Entering new markets and introducing new products often implies changes in performance requirements for the different business processes. Change to more tailor-made products requires a better flexibility, better skilled operators, new performance indicators, more responsive production management systems, etc. However, business has to continue as usual with existing customers.

6.2.4 Integration of performance requirements

Definition of performance requirements is a matter of integrating the set of stakeholder requirements and strategic requirements. These requirements should cover all important aspects of competitiveness. Performance requirements could then be defined as:

> *Qualities that a company is expected to possess to fulfill the needs of present and future stakeholders in order to sustain or enhance competitiveness.*

In most cases, a stakeholder solely focuses on ability to fulfill needs. Lacking qualities could be compensated by other qualities, but having the right qualities is assumed to be the most cost-effective strategy. However, there will probably exist several conflicts of interest. A simple summation of the requirements is then impossible. A trade-off is required to solve these problems, and it is defined as:

> *A compromise between conflicting interests with the need to maintain equilibrium.*

Revealing trade-offs are important because hidden trade-offs are potential sources for dissatisfaction. This process will provide valuable information about the nature of competitiveness. It is often said that the planning process is more important than the plan itself.

What we are really looking for, is to develop a common understanding of the priorities within the company. However, priorities may vary within the company. In this book we present a simplified model by looking at the company as a homogenous unit. By allocating the interfaces to the different business processes, this methodology could easily be extended to support business process modeling. The process of matching external performance

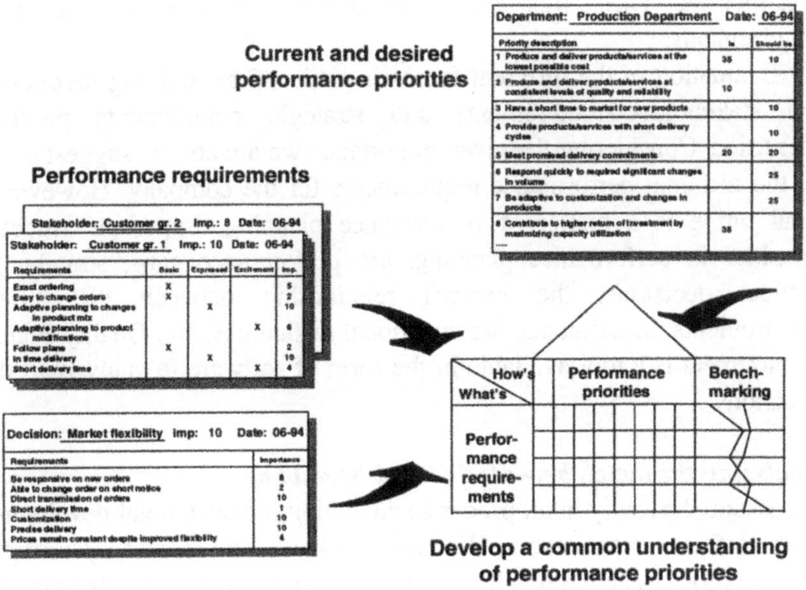

Figure 6.15 A model for development of performance priorities.

requirements with current and desired priorities is shown in Figure 6.15. Integration between market strategies and performance strategies could be obtained by applying the methodology for different market segments.

Development of a common understanding of priorities is a process that includes development of the organizational culture. People within an organization have values and beliefs that are difficult to change. Simply stating new priorities is not sufficient. Three tasks could be identified:

1. Identify the performance drivers for the business
2. Analyze the match between 'is' and 'should be'
3. Develop a common understanding of priorities

Identify the performance drivers for the business

Performance planning has to be based upon the real driving forces of the business. Suggestions of performance drivers could be based on experience or provided by creative approaches like brainstorming. Surveys are also valuable sources for input. It is important to be open-minded and not include evaluation in this process to cover all possibilities. Awareness of the influence from existing values and beliefs is important to avoid ending up along old tracks. The inertia in changing values and beliefs is considerable. There often exists a mismatch between values and the real requirements.

Evaluation of the different forces is achieved by auditing them against the stakeholder requirements and strategic requirements previously identified. Considering the given importance we are able to suggest a rating of the different performance requirements for the company. However, the final rating is a matter of performance planning. Important aspects to consider in performance planning, are performance gaps, strategic and tactical decisions, the internal relationship between the different requirements, investments, organizational difficulties, etc. Quality Function Deployment is a tool available in the form of software to analyze all these decisions.

Analyzing the match between 'is' and 'should be'

An essential activity in all processes involving organizational development, is to define current status. We have to know the existing priorities to choose appropriate approaches to change. An internal survey including the different departments and organizational levels could reveal the existing set of priorities. Lynch and Cross (1991) suggest an approach where each department manager indicates existing priorities. This is done by distributing a given number of points on the different predefined priorities from the previous step. Simple forms could be used to perform this gap analysis concerning both current status and desired changes in priorities. Figure 6.16 shows an example of a form for a departmental survey.

Figure 6.16 shows a development from a focus on cost efficiency towards flexibility. This change is a result of an adjustment to the stakeholder and strategic requirements. We do not neglect costs, but realize that costs are related to activities. Flexibility is a necessary quality if the relationship between income and expenses and hence profitability should be maintained in a changing environment.

A systematization of all forms from the complete survey and comparison with the previously defined priorities provides answers on two important questions:

- Degree of convergence or divergence among the different functions
- Size of the gap between 'is' and 'should be'

Develop a common understanding of priorities

A common understanding of priorities is necessary to develop an efficient performance plan and to achieve the best possible integration with the

Department: Production Department		Date: 06-94
Priority description	Is	Should be
1 Produce and deliver products/services at the lowest possible cost	35	10
2 Produce and deliver products/services at consistent levels of quality and reliability	10	
3 Have a short time to market for new products		10
4 Provide products/services with short delivery cycles		10
5 Meet promised delivery commitments	20	20
6 Respond quickly to required significant changes in volume		25
7 Be adaptive to customization and changes in products		25
8 Contribute to higher return of investment by maximizing capacity utilization	35	

Figure 6.16 Example of a gap analysis form.

business strategy. This is a very difficult exercise, because existing paradigms and values are challenged. Managers from different departments seldom agree upon how to respond to the changing environment. The performance planning model provides some important characteristics that will ease this process:

- Involves all employees based on their qualifications
- Visual and easy to apply
- Boosts the existing functional 'silos'
- External view that pushes the discussion above internal prestige
- Try-and-fail approach that makes 'simulation' possible
- Forces the important questions on the agenda

Applying the model is no guarantee for acceptance of common priorities, but these elements amount to a quantum a leap compared to traditional planning methods. Traditional methods are often too specialized to achieve real involvement of those responsible for implementing the plans. Involvement is perhaps the most important factor in achieving acceptance by all affected parties. Understanding and acceptance of the performance priorities are absolutely necessary for development of a consistent set of action plans throughout the organization that will result in real business achievements. Otherwise, contradicting efforts may cause a loss in efficiency and hence competitive position.

Tools like Nominal Group Technique (NGT) could be applied in the process to achieve a common understanding. It is important to look at this as a part of organizational development. Performance planning is only one of the activities within performance management, and they have to be seen as a whole. Training and information are essential to benefit from the tools and the methods described.

6.3 QUALITY FUNCTION DEPLOYMENT AS A PLANNING TOOL

The scope of Quality Function Deployment (QFD) is a customer-oriented approach to product innovation. It is a methodology to structure needs, expectations and requirements of the customers and then turn these into detailed product and process requirements. Even if QFD was originally developed as a tool for the product development process, the methodology can be used on several other problems.

6.3.1 The theory behind QFD

The ideas behind QFD go back to the late 1960s as we know the technique today (Akao, 1990), but the first documentation of QFD was written by Akao in 1972. In the same year, the QFD technique was introduced at Mitsubishi Kobe Shipyard in Japan, and Toyota adopted the methodology in their product development process five years later (Gevirtz, 1994). In 1978, Akao and Mizuno systematized the basic ideas and concepts of QFD that were used in the Japanese industry so far.

The basic structure of QFD

To ensure that the customer voice is deployed through the whole development process, it is essential that the process is based on the customer viewpoint.

Usually, we can discern four phases in the product development process:

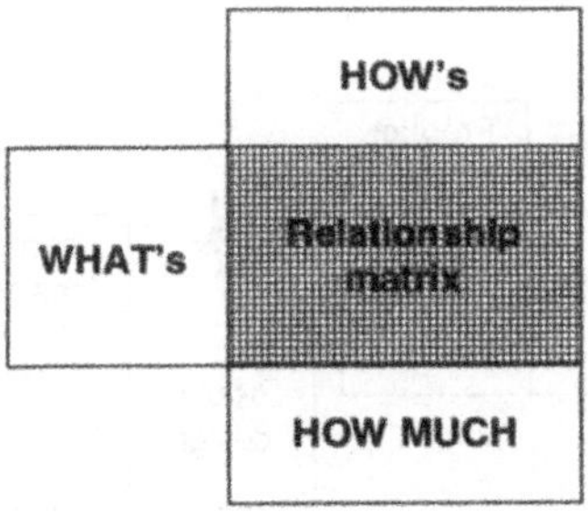

Figure 6.17 Basic structure of QFD.

- Customer requirements to product concepts
- Product concepts to product design
- Product design to process design
- Process design to production operations

Each step in the development process must be consistent with the originally customer needs and expectations. The basic structure of QFD is the relation matrix between requirements of the different levels in the process as shown by the chart in Figure 6.17.

WHAT's is meaning what to obtain which represents the target goals of the analysis. In the first stage of the development process this equates to the customer requirements. *HOW's* is the means of how to obtain the targets and likewise corresponds to the product concepts' requirements. When taken into account the importance of each goal and analyzing how each of the means is supporting the goals, the importance and the required performance level of each mean can be determined. This is represented by *HOW MUCH* in the chart and corresponds to the level of customer satisfaction or importance and decision basis for different concepts.

Each phase in the product development process can be linked together as a chain of relation matrices that ensure a continuous deployment of the customer needs through the development process. This makes a cascade of QFD charts as shown in Figure 6.18.

'House of Quality' and benchmarking

The relation matrix makes the basis of the QFD methodology, but several other features that are important for the decision process can be added to

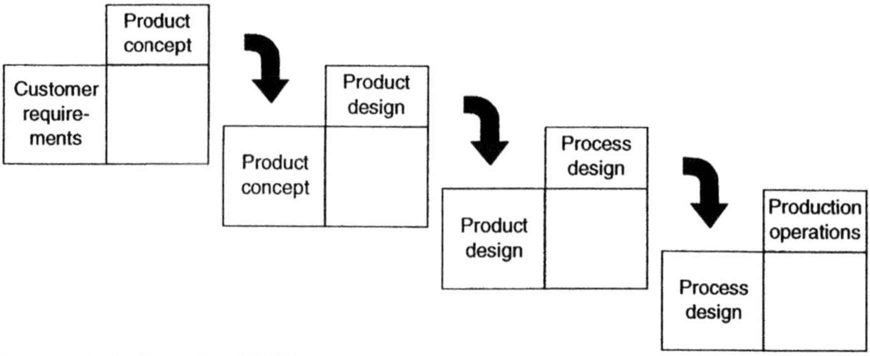

Figure 6.18 Cascade of QFD charts.

the matrices. Figure 6.19 is showing a basic ‘House of Quality’ with ‘rooms’ of the various steps in the QFD process (Govers, 1994).

The QFD-process is carried out by complementing the different ‘rooms’ in the ‘House of Quality’. First step is to determine the *WHAT’s* that is the description of goals to obtain and the importance rating of each item.

The questions *WHY* focus on which requirements the company has to fulfill. Competitive benchmarking is an important tool to derive areas for improvements. The competitiveness is illustrated by comparing how the other competitors match the customer requirements in proportion to our product. An example of illustrating the *SCORES* of the different products is shown in Figure 6.20.

After finding the means of *HOW* the goals should be obtained, the

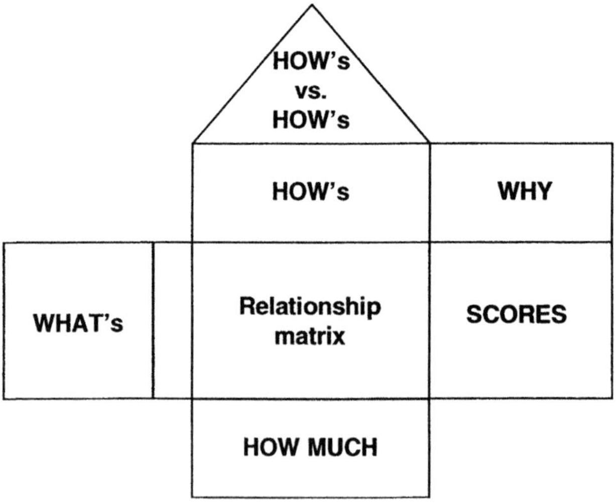

Figure 6.19 House of Quality.

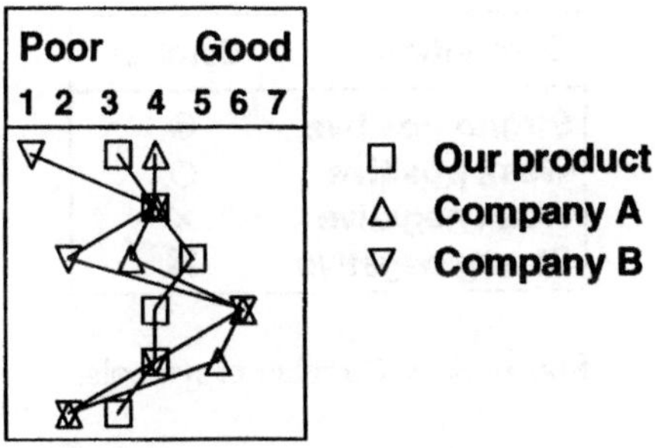

Figure 6.20 Competitive benchmarking.

relationship matrix between the *HOW's* and *WHAT's*, can be worked out. The relationships can have both signs, but the means are usually defined in such a way that they give positive influence on the goals. It is often favorable to use a limited number of relationship levels to make the chart more easy to read. Commonly used symbols and examples of weighting factors describing the level of relationship, are given in Figure 6.21.

The relation matrix can be based on several conditions. Often it is not possible to detect a direct relationship between the factors. Statistical analysis of historical data or knowledge about similar relationships can provide important support.

Targets of *HOW's* determine the *HOW MUCH's*. To be able to determine the performance of the *HOW's*, it is importance that the targets are as measurable or descriptive as possible. To compare our performance with the most important competitors, the tool provides benchmarking of the *HOW's* too.

The 'roof' of the 'House of Quality' contains the correlation matrix between the different *HOW's*. This matrix indicates both relations and conflicts. Specially, the conflicts are very important to detect because they represent conditions which determine trade-offs. Figure 6.22 shows commonly-used signs for correlation factors.

The importance rating of each *HOW's* can be calculated as the sum of

Relationship	Symbol	Weighting Factor
Weak	△	1
Medium	○	3
Strong	◉	9

Figure 6.21 Commonly used symbols and weighting factors for relationships.

Correlation	Symbol
Strong positive	◉
Weak positive	○
Weak negative	×
Strong negative	⋇

Figure 6.22 Correlation symbols.

products of the importance ratings of the *WHAT's* and the weighted relationship factors. A *HOW* with strong relationship to many important *WHAT's* will obtain a high important ranking and should be preferred with relationship to non-important goals. The decision basis can be improved by considering other factors that have influence on the decision process, by multiplying weights to the importance rating to calculate new ratings. Examples are implementation difficulty, probability to success and frequency and severity of problems to solve. The ratings are also useful as input if the requirements have to be deployed further.

The ratings are useful to compare different *HOW's*, but it is important not to trust blindly on these numbers. Even if an item gets ten times higher rating than another one, it is not sure that it is ten times more important. The calculations in the chart are sensitive for errors and selection of weights. However, if we can accept the ratings and there is a trade-off between the two items, the item with the highest rating should be preferred.

QFD as a general planning tool

Even if the QFD methodology is developed to improve the product development process, the techniques can be applied on many other problems. The technique is useful to make sure that the overall strategy is deployed through a decision process. It is a general approach to analyze any relationships that can be represented by *WHAT's* and *HOW's*. The methodology can also be applied in a performance planning process.

6.3.2 An example of performance planning with QFD

The QFD methodology offers a favorable tool for transforming stakeholder and strategic requirements into performance requirements for a given business unit. The methodology can be applied for any areas of the business such as the business process, organizational units and limited problem areas. In the following example, we have applied the methodology to develop performance requirements for production management using the QFD Designer software from Qualisoft. Figure 6.23 shows the 'House of Quality' for the relationship between some performance parameters for production management and a sample of stakeholder and strategic requirements. The methodology guides us through the process of defining requirements to the performance parameters.

The stakeholder and strategic requirements and related importance can be presented directly in the QFD chart as *WHAT's*. All available sources for evaluation of performance of each requirement are important input to identify performance gaps and areas for improvement. This information is presented as *WHY's*. Input could be metrics or evaluations. Performance gaps could be identified by customer requirements, customer surveys, competitive benchmarking, external audits, etc.

In the QFD-chart in Figure 6.23 we have included a customer survey where we have compared the company's scores with two competitors. We have also indicated a desired target level to define the performance gap. The chart indicates that conformance quality is both very important and has a large performance gap. Therefore, we should be aware of improvements of performance parameters that influence on conformance quality.

Next step is to develop a set of possible performance parameters that forms the *HOW's*. We have chosen parameters from production management. Each possible relationship between the stakeholder and strategic requirements and the performance parameters of production management is then investigated and the characteristics of the relationship are evaluated in the relation matrix.

Several methods exist to determine the relationships. Some relations are obvious, but others can be difficult to recognize. Statistical analysis and analysis of a business model to extract the relations, are helpful tools.

The QFD chart exposes two empty lines in the relation matrix concerning suppliers. This indicates that improvement of performance parameters of production management has little influence on the relationship with suppliers. Satisfaction of supplier relationship must be covered by other business functions.

The targets of the performance parameter must be as measurable or descriptive as possible. To be a follow-up target, it is important that adequate measuring can be done. However, in the QFD chart, we are not able to find measurable targets for all performance parameters, and further division of the performance parameter is necessary if we want to find targets to follow up these parameters.

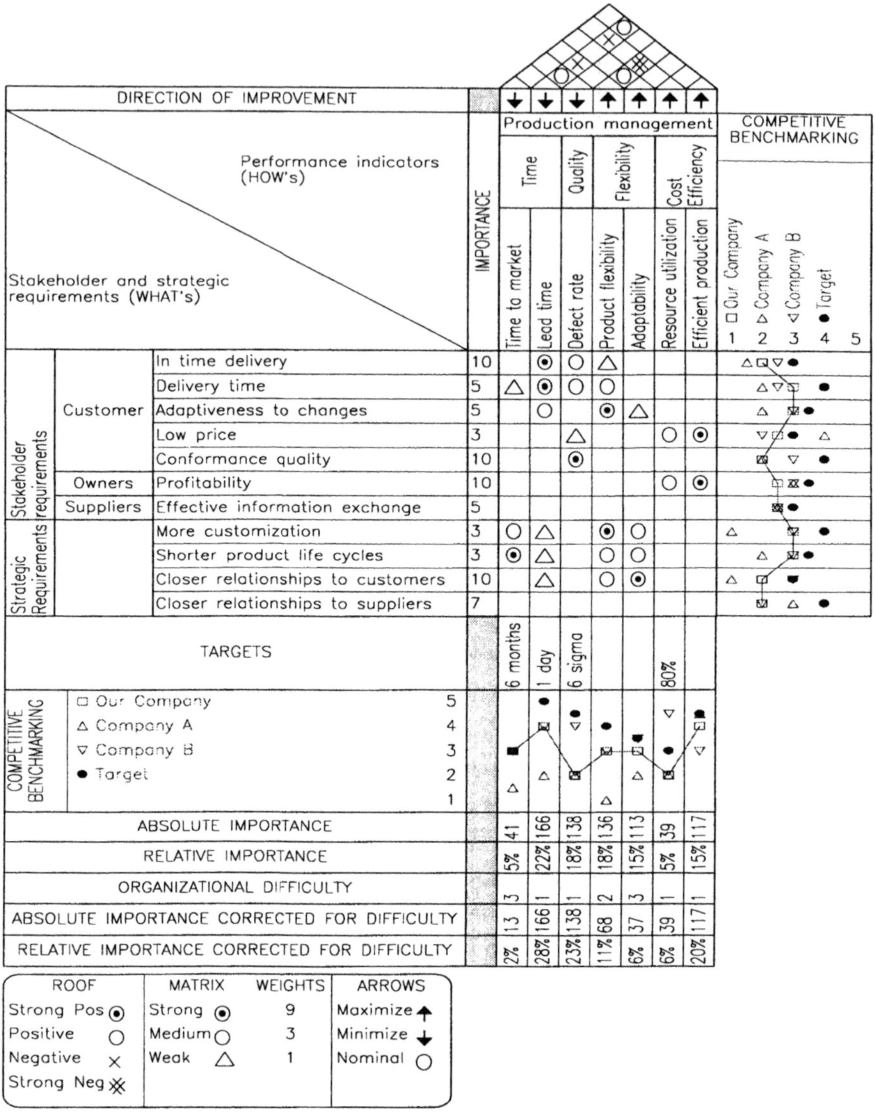

Figure 6.23 An example of a 'House of Quality' for production management.

Comparison of the performance level with other companies provides an effective tool for recognition of the performance gap and establishment of possible targets. The QFD chart shows that the defect level has high influence on conformance quality and a high performance gap. Therefore, reducing the defect level seems to be a reasonable strategy.

The roof of the 'House of Quality' indicates trade-offs and possibilities for concurrent improvements of the performance parameters. The QFD chart shows that more efficient production and diminution of defect level have a positive influence on improving the lead times. Likewise, increase of flexibility in production also involves the production is more adaptation for future changes in environment. On the other hand, there is a trade-off between resource utilization and both flexibility and lead time.

The importance rating of the different performance parameters is calculated by the use of weights for the relationship and importance of the different stakeholder and strategic requirements. The QFD chart shows both absolute and relative numbers.

It is obvious that some performance parameters are easier to improve than others. This can be due to different implementation problems, knowledge, organizational obstacles and so forth. In the QFD chart we have also calculated the importance rating when taking into account different organizational difficulties. The results show that improvement of lead time, defect rate and more efficient production, are the main means for improving the stakeholder and strategic requirements. From the 'roof', we also remember that improvement of these factors can gain advantages of concurrent improvement at the sacrifice of resource utilization.

The ability to develop strategies based on performance requirements have been tested on the TOPP database (Bredrup and Bredrup, 1994b). QFD was then used to analyze the results of different strategies concerning time, cost, quality and flexibility. All strategies were then compared with different performance indicators with relative importance given by the companies. The success of various strategies was then evaluated. The findings from this analysis could be summarized in:

- Quality management is given high importance, but fail to support performance indicators due to poor fulfillment
- Speeding up the enterprise has a high potential of performance improvements
- The enterprises are highly cost focused, even if it is not said to be important
- The enterprises fail to utilize and gain advantage from high flexibility

However, the main conclusion from this analysis is that QFD is an appropriate tool for designing strategies that fulfill a complex set of performance requirements. It helps to visualize trade-offs and force the company to develop a coherent strategy.

PART THREE

Performance Review

7. MODELING IN MANUFACTURING ENTERPRISES

Models in the sense used in this book, comprises a visualization of a situation that either belongs to the past, presently exists or we predict to happen sometime in the future. A model can be a figure, a flow-diagram, a program, prototype, etc. How the model is built depends on the purpose of making the model. For purposes of making models of manufacturing aspects, we are limited only by our imagination. For large companies and complex situations, modeling is both a major task and a way to support understanding of complex cases.

For manufacturers, modeling can be a tool for analyzing or simulating, for resource planning, performance measurement, designing products, process management and so on. The core of building models is that subsequently some action should be taken, see Figure 7.1. For instance, if we made a model of the procurement of subcomponents, we could make a model of the procurement process of today. Then we can visualize the difference of the wanted situation by a model of tomorrow's procurement process. Between these two situations we have to implement some changes to achieve the wanted state of nature. Again, these processes of change can themselves be modeled to be sure that the right factor of influence is included and thereby the objective is achieved.

For all kinds of requirements, we can try to predict the future, always having in mind that planning is done with some degree of uncertainty. On the other hand a model can not comprise a total situation. It is not feasible to include all kinds of elements for the model to be 100% accurate. A model too complex to make simple changes, is not an appropriate model. In

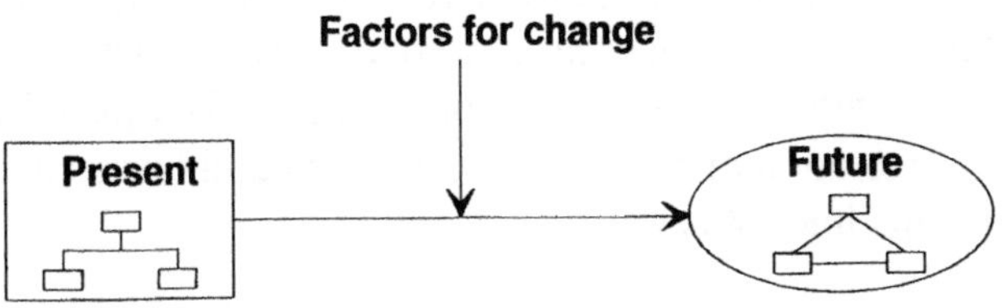

Figure 7.1 Model of present and future situation for inducing change.

this chapter we will see some of the current modeling aspects of manufacturing. We build a model of the entire company out of the basic modeling approaches of manufacturing, that is the product, process and organization model.

7.1 MODELING THE COMPANY FOR IMPROVEMENT

Modeling could serve different purposes:

- manufacturing aspects (layout, resource planning, bottlenecks, product, process, organization);
- develop an understanding of the enterprise operations to identify sources for improvement;
- describe the system and its interfaces to make measurement possible.

A close link between the model and improvement tools, and between the model and measurement methods is necessary. These links will give rapid and reliable input to the process of improvement. Through a model the new results can be monitored and evaluated, and corrective actions can be realized if necessary.

A shift in business such as a shift in managing manufacturing companies leads to a new view of a conglomerate company. Managing the cross-functional processes can give increased motivation to find new ways to improvement. The companies need new types of managers to handle this change. They have to be broad-minded and have an inclusive attitude to the activities going on in the company. The managers are responsible for encouraging people and departments to think broadly and unselfish. The objective is to create a better environment internally for cooperation and externally for an attractive partner, producer and supplier.

In this chapter we will present some of the approaches to modeling various aspects of manufacturing. Modeling aims at trying to set the factors of manufacturing in one model that can be used to give an overview of the situation and for future changes for improvement. The planning aspect is taken into consideration of creating models of what the current situation is, and to use this in either of two ways:

1. Analyze the model to state any out-of-specifications conditions or faults.

2. Create a model of the wanted future situation and use it later for evaluation of results achieved.

We present shortly two concepts that are influencing many companies, and, in fact ours, in the development of a balanced business modeling method. They serve different purposes and are seldom used in isolation on the operational level.

7.2 MODELING OF MANUFACTURING ASPECTS

Modeling the manufacturing processes is necessary to support:

- planning of manufacturing activities (defined by the product's life cycle);
- re-organization of labor and machines;
- financial allocation.

Collection and organization of technical information from all functions involved in generating products is one of the major tasks in building models (Prytz, 1994). The ultimate objective is to create an integrated model to link the life cycle stages of the product. The results of production actions can be visualized through a Virtual Manufacturing Environment (VME). We will come back to VME later in this chapter.

Most companies have a number of objectives they want to achieve. These are shortterm, like each day answering all incoming questions from customers, and longterm as for instance in being a cornerstone company employing and sustaining the local area. Objectives can be split up until the objectives consist of operational activities, single or dependent. This is necessary to be able to accomplish the objectives.

> The Company, Inc., has decided that they need to focus on customer satisfaction. They produce valves. To expand their market they have decided to sell valves to several types of customers. According to their new objective they start to use a House of Quality or Quality Function Deployment (QFD) to improve their communication with their customers. They want this to help them to achieve satisfied customers, through customer specific valves. The house of quality can be used as a model integrating both the customer requirements and the Company's ability to meet these requirements.

Organization structure has traditionally been the common business model. The organization structure does not enable the organization to focus on the customer. Organizational focus has led to departmentalisation and specialisation of activities and a diversity between the functions in a company. Research has shown that the specialisation paradigm provides less productivity gain than more modern approaches (Istvan, 1992). So, the challenge for companies today is to manage processes across the functions. The organizational structure within a company will still be necessary, but a new structure is becoming strategically important for the new philosophies to be realized.

7.2.1 Product modeling

We define a product model as (IMS Globeman 21, 1994):

> *A product model is a generic model used for representing all types of artefacts which appear in the course of manufacturing. It represents target products, their materials and intermediate products, tools and machines, and any other manufacturing resources and environmental objects.*

For flexible manufacturing, exposed to international competition, a product model should respond to the following demands:

- globalization of manufacturing activities in a decentralised environment;
- elimination of isolated islands of automation in production;
- flexibility to customer needs;
- creation of a manufacturing process integrated with the organization to improve man/machine interaction.

A product model is often a geometric model. The development of object-oriented modeling techniques in recent years have proved to give a considerable contribution to productivity when developing geometric models (Bjørke and Myklebust, 1992).

With the change toward more global competition followed with global manufacturing, great focus has been put on international collaboration, and in particular creation of international standards. Within the product modeling area, ISO/STEP has been raised. STEP is the Standard for the Exchange of Product Model Data. It is developed under the ISO (International Standard Organization), and the work on this standard started in 1984 (Owen, 1993). The objective of the standard is to standardize the exchange of product model data. To realize this, the representation of the product geometry and characteristics must be with reference to a standard. The standard will give great contribution to improved cooperation between design and production department.

We will not go in to details on product modeling, but emphasize that product models must be integrated with other models of the company. This integration is also important to include the manufacturing processes to make the design of products adequate for manufacturing. The same consideration should be done to include customer demands for the product, and try to link various aspects of the product into a model. This brings us to a new approach called Conceptual Engineering.

Conceptual Engineering

The most important decisions that a designer makes, are related to the principles or concepts that are used for the design. That these decisions are important, can be illustrated by the obvious consequences of a decision to go for a car body design based on a space frame and panels as opposed to a unitised design. This decision alone would have far greater impact on the properties of the final product (in terms of cost, production, transport, destruction and recycling) than any amount of detain design work. These concepts are of a different nature than the geometry's and structures that are typical of detail design, and they appear at all levels of a product (in a car you would have a body concept, a seat concept, a bumper concept, an engine concept, a fuel flow regulation concept and so on).

A number of different models exist. They differ in their description of the conceptual and the relationship between the conceptual and the detailed models are not a fully understood area. One of the most promising approaches, distinguish between four different areas of product data. These are:

1. the process area
2. the function
3. the organ area
4. the construction area.

The areas are casually linked so that the function represents what the product should do, the organ represents how it does it in principle, and the construction entities how these are materialised. The process represents a transformation view of the product. This model is probably the most advanced 'total' product modeling approach known today. Conceptual design is first of all a question of generating alternative solutions, evaluating them against requirements and each other, and select one alternative. Conceptual design and design models have been reported to give a complete and superior approach to the management of large and complex projects like offshore installations. The design models must be developed in close relation to the conceptual engineering evolvement to bring the new approach a closer step to reality.

7.2.2 Process modeling in manufacturing

Models of the production systems are made to visualize the influencing factors of the production. These factors can be resources coming in, products and services going out, the processes ongoing in the factory, etc. One of the basic and acknowledged ones is the Walrasian model, see Figure 7.2.

The Walrasian model depicts the transformation process of production factors into finished products (Rolstadås, 1991).

The Walrasian model can represent both functions and cross-functional processes. Horizontally it represents the functions in the company. Where the cross surrounded by a circle is indicated, the process or activity is utilized by a function. The degree of technical dependency is given by a technical coefficient. The model does not include ordering of materials or component, nor includes production operation (routing) as in real

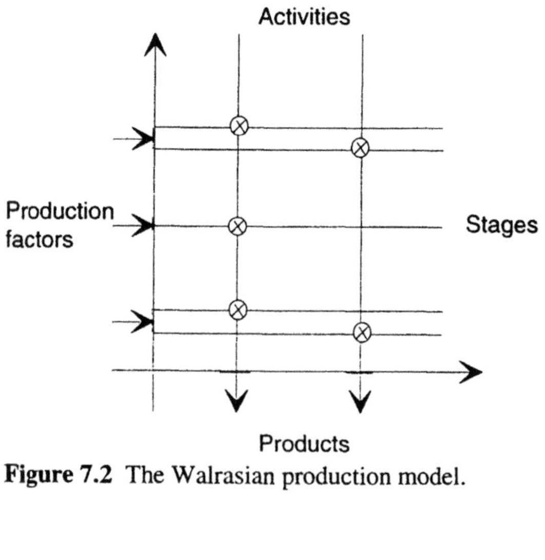

Figure 7.2 The Walrasian production model.

production systems. Real production systems are also in a dynamic environment, so in addition to the static Walrasian model other dimensions should be considered. The model is often extended to include also:

- ordering of products
- required resources
- a time dimension.

Process modeling in the factory is, of course, concerned with planning to execute the manufacturing processes. Process modeling is concerned with the following approaches (Bjørke and Myklebust, 1992):

- process planning
- operation planning
- NC planning
- material and resource planning.

This means the merging of open information systems with other systems keeping product information, with the objective to achieve a quicker and safer method for preparing the production data. The type of product that is manufactured lays the requirements for how advanced a process planning system must be. For complex products and very customer-specific products, efficient process modeling solutions can be hard to find.

Process models are used for representing all the physical processes that are required for representing product behavior and manufacturing processes. In principle, they are generic models, but their primary aim for virtual manufacturing is to predict unknown product behavior and to evaluate ability to produce in unusual production processes. A flexible manufacturing system (FMS) is necessary to handle the changes occurring from new evaluation done in the model. FMS and virtual manufacturing will make manufacturers capable of meeting ever-changing customer demands. This will be in accordance with a TQM philosophy to meet the customer demands. So a movement toward this kind of manufacturing will be vital for survival of international competing enterprises.

Today's industries require to improve their manufacturing and business processes to assure a better and faster response to their customer's requirements, higher quality for their products, and more flexibility and faster response in the introduction of new products.

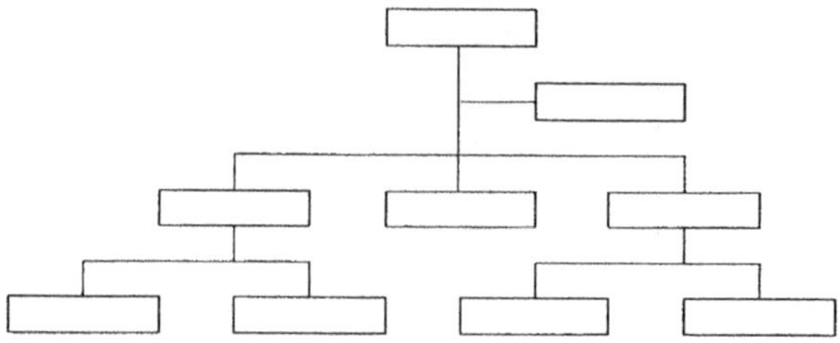

Figure 7.3 An organization structure.

To achieve these goals, many kinds of technology have been pursued in each specific domain. This includes not only in factory production, but also in various types of processes such as product design, manufacturing design, logistics preparation, tactical decision, and high level strategic decision.

7.2.3 Models of organizations

If you ask someone to show you a model of their company, in most cases you will be presented an organization chart. Organization charts have traditionally been the visible picture of the company.

Even today many companies have a rigid hierarchical structure, see Figure 7.3. Models like this show only the rigid and stabile part of the company. For the employees, these pictures of their employer give no indication or motivation of cross-functional cooperation in the company. An organization chart can only be used to determine where manager's responsibility lies, and for the employee who their manager(s) is. For the future evolvement of the company, and regarding the changes in market and business, companies must develop more advanced models to include other dimensions for versatile relations. These relations can be between people, internal and external communication, organizational factors influencing production, a balanced organization with respect to competence and linking the company with its customers and suppliers.

7.3 PRODUCT MODELING

Here we show two examples of product modeling put into practise in manufacturing companies today.

7.3.1 Concurrent engineering

Concurrent engineering has been developed for several years now, and has only recently been familiar to the manufacturing companies. Especially enterprises working with large-scale projects try to implement the factors of concurrent engineering. What then is concurrent engineering?

An overall objective for working concurrently is lead time reduction in development of a product. If a new product demands a project with several contractors and suppliers, then these can not sit and wait for the other partners in the project to finish their work. The solution for concurrent engineering is that they have to work concurrently, in parallel. This requires communication between the partners through the entire work process.

The other objective for concurrent engineering is that with parallel working the specifications of a product can be influenced by all the functions working with the product. In other words the entire product life cycle can be integrated in the development of a product. These are concerns like user demands, recycling, demolition, maintenance, etc. This is why concurrent engineering also is refereed to as simultaneous engineering. Including a life cycle strategy from the early beginning will improve the quality of the product by take into consideration future demands for the product. The integration of cross-functional or inter-departmental cooperation is the foundation for concurrent engineering. The traditional organization structure is that of very limited communication between design and manufacturing department. This creates a lack of understanding for the processes in the chain of the products life cycle. What happens at the interface points between the departments, is that production has a hard time producing the product because of some design choices made by engineering. These choices are often made arbitrarily, and they could have been slightly changed to dramatically ease the production task.

We conclude that concurrent engineering has two main objectives:

- lead time reduction;
- improved quality through the entire life cycle due to cross-functional synergy.

A successful implementation of concurrent engineering can result in both reduced development time and improved quality. Companies have already recognised this, and are striving to have the entire company working according to the principles of concurrent engineering. To make such an implementation successful the companies almost always need computer tools to process all the information and data that is required when a large-scale project is running. If all the contractors and suppliers use the same practises, like the same tools, standards, etc., the companies can be connected with each other and cooperate on-line and much closer. When the companies are communicating they want to use their workstations were they have all the information and data stored. This way of communicating has exploded in the last few years.

It is presently only isolated systems that take the concurrent engineering objectives into account. The lack of a system that takes the entire life cycle into consideration makes it necessary to research systems that can link the already existing systems together.

Here are some of the tools/methodologies that have to be developed before the entire life cycle approach can be implemented:

- method for incorporation of life cycle issues and costs in the design phase;
- early involvement of subcontractors and suppliers to give input to design specifications;
- material flow analysis in all life cycle phases;
- emission risk models to prevent any disasters or damages to both humans and environment;
- information stored for future use;
- risk analysis (accidents, fire, explosions, chemicals, materials, energy);
- optimization of all resources (materials and energy) in the entire life cycle;
- computerised system to support concurrent engineering that should have an interface to other management and CAD systems, have the possibility to design despite lack of data and information, provide easy handling of design changes, and have a good user interface.

One of the approaches, has been Design For Manufacturing (DFM). It aims at reducing the time and cost used for manufacturing. At the same time it accomplishes the quality that is specified. This is done through reduction of the number of components, reducing the number of assembly operations, doing the things right the first time, pre-evaluation of the value of the product, high reliability processes.

There have already been some approaches to create a tool for implementing all the life cycle phases in the development of a new product or the design for manufacturing. The best known is the Design For Assembly (DFA). DFA is a program that takes manufacturing issues into consideration already in the early design stages. The program consists of one part were the designer specifies the design of the product and the program answers with the key assembly numbers like assembly efficiency, breakdowns, etc. This and quite similar programs that exists at present, make it possible to foresee the next steps in the product's life cycle. This kind of program must be utilized through the life cycle stages.

The difficulties have often been that the design and the production departments have had little or no communication. This prevents the concurrent engineering approach from working appropriately. According to Rembold *et al.*, (1993) this can be improved if:

- the designer is trained in manufacturing principles;
- the manufacturing engineer learns design engineering;
- close cooperation exists between the design and the manufacturing engineer during the design of a product.

7.3.2 Rapid prototyping

Because of the need to reduce lead time, fabrication costs, improve efficiency, and increase flexibility without sacrificing quality, rapid prototyping has become a remedy for especially the design stage. Its virtue lies in the words. Rapid meaning fast and quick, and prototyping bringing forward a pre-manufacturing solid model of the product. An early prototype of the model can have great impact on the selection of materials, manufacturing methods, design changes, etc. If the process of making one prototype is expensive and/or complex, then only a few selected parts are made in advance of the production setup. The objectives of rapid prototyping (RP) are therefore fast and accurate production of one or few pre-manufacturing prototypes.

Present practise is stereolithography, but other methods are also in use such as selective laser sintering (Atwood *et al.*, 1993). Stereolithography also called layer manufacturing, is a method where complex plastic parts are produced from computer-generated data. The data can come from a 3D CAD model directly linked to the stereolithography machine. When the production-like prototype is ready for fabrication, the proper method of manufacturing is necessary to determine. A CAD model datafile can be translated into a manufacturing datafile and thereby generate tool paths. Rapid prototyping implies that more design engineers are beginning to use solid modeling as a design tool.

We assume that the future worker will be more mobile and will rely on a 24-hour global electronic, satellite and wireless technological infrastructure. This will give the worker undreamed-of possibilities to operate in a global network. The result might presumably be more and larger alliances across the regions and between businesses that we see working separately today.

Work is always done under time pressures. But in today's paradigm, the need to manage time has become the critical priority. Time-to-market, time-to-manufacture, just-in-time, and dozens of similar time-based banners make it very clear how important control over time has become. Given the emphasis on using time-to-market as a competitive advantage, we would expect organizations to look for more ways to reduce time they take to turn out their products and services.

7.4 PROCESS MODELING

Process modeling is today linked to the term and area of re-engineering. Here process modeling is discussed as a part of the re-engineering process, but also in the aspect of Quality Awards. Quality Awards is one approach to consider in terms of processes, and it can be used very positively for many companies.

7.4.1 Business process modeling and re-engineering

Central in the new productivity paradigm is the term processes. Numerous voices raise to encourage companies to 'think processes' instead of functions. Earlier it was called procedures, but today we think of processes that go horizontally through the company. Integration of process thinking in companies is first of all for inducing improvement and measurement.

Traditional functional modeling methods fail to measure and support improvement in cross-functional activities. Traditional functional processes include project management tools like for example, Gantt-diagrams. A Gantt-diagram can not depict the complexity and dependency between

Figure 7.4 Definition of business process.

functions and processes. The traditional models are also focusing on traditional measurement as cost and time. The models are not sufficiently flexible to include the new focuses on quality, flexibility and improvement. As complexity grows these cross-functional activities increase in number and importance, and greater becomes the challenge to handle all the influencing factors optimally.

To look at the processes instead of the functions, has in recent years been described as business process redesign, business process re-engineering, business process innovation, apart from several nicknames. They all aim at the same objective of improving the processes across the traditional functions that comprise a company. A discussion of whether business process re-engineering is small incremental improvements or large innovative shake-ups of the company is going on. We will not take up the discussion here, but recognise that business process re-engineering can be used for both purposes. This is dependent on how we define it and what type of objectives we have for going through the change. A process orientation to business involves elements of structure, focus, measurement, ownership and customers, according to Davenport (1993). Let's see what processes really are in manufacturing firms.

A business process could be defined as (see also Figure 7.4):

> *Work activities that use material, information, equipment and people to provide output and service.*

When processes are structured and clearly defined, measuring their performance, inputs, and outputs is possible. The argument is the same as for continuous improvement in the quality term. There has to be someone responsible for the process, someone who knows the process, knows what influences its performance and what the customers of the process want. The people responsible have to be able to have control over the influencing factors, and for this they need training.

For all kinds of companies the selection of main business processes is a process of clarifying what the companies major strengths are. Out of this a business platform can be created. Processes in manufacturing company's are (Davenport, 1993):

Operational:

- product development
- customer acquisition
- customer requirements identification
- manufacturing
- integrated logistics
- order management
- post-sales service

Management:

- performance monitoring
- information management
- asset management
- human resource management
- planning and resource allocation

Here we introduce one extra type of process that includes a new dimension, or a cybernetic loop, the cycle process.

Cycles:

- equipment life cycle management
- human resources life cycle management
- product life cycle management
- cash flow management
- material flow management

Operating processes are identical to primary activities in Porter's value chain, whereas management processes are identical to Porter's support activities. Cycle is a newcomer. Most aspects concerning cycles are covered by other processes, but some important aspects are left out in other models. Activities appearing in more than one process are no problem, because the processes are independent of the organizational structure. The main purpose of processes is to improve the performance across the borders of functions.

When thinking in terms of processes we have to focus on the few important ones, rather than all processes. Process thinking is a radical change for many companies and businesses, and it requires a change for all employees toward realizing the customer/client chain through the entire company.

There exists no universal business process modeling method that is optimal for every company or industry. There are certainly consultants trying to advocate the opposite, but the purpose here is to promote the understanding of business processes and not to sell a defined concept. Each company has to design its own business processes to be able to gain enough knowledge to improve them. Designing the business processes implicates several difficult decisions and matters of prioritizing. If this is done externally or is even defined in advance according to a particular method, the necessary organizational development may fail to appear. The need for organizational development should never be under-valued else the introduction of process thinking is doomed to fail due to resistance within the organization.

When modeling, a company has to define the purpose and the environment they are facing. The stakeholder model is of special importance regarding the evaluation of the output.

7.4.2 Quality awards criteria as a way to model the business processes

Recent years focus on Total Quality Management has promoted quality awards for acknowledgement of results in quality work in manufacturing, as well as service companies. Quality awards like the American Malcolm Baldrige Award, the Japanese Deming Prize and the European Quality Award have influenced behavior in many companies. Several companies have adapted the criteria as a way to model their business. Although only a limited number of companies actually apply for the award, their influence on industrial behavior is significant. Quality aspects are naturally the main focus of quality awards criteria, but they are useful to understand market relations as well. The Baldrige Award has the following seven broad groups of criteria (relative importance is given as number of points):

- leadership (100 points);

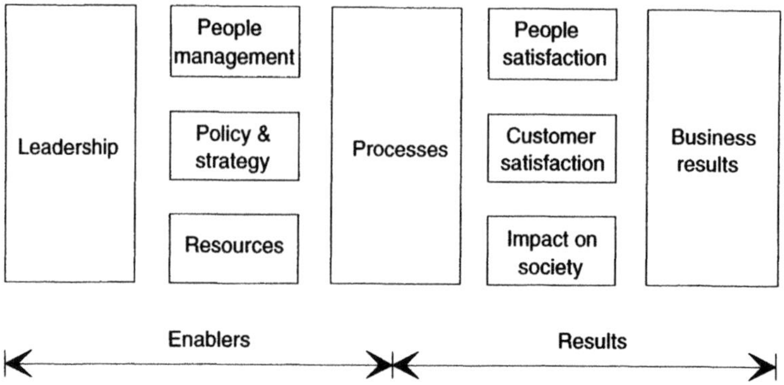

Figure 7.5 The European Quality Award as a business model.

- information and analysis (70 points);
- strategic quality planning (60 points);
- human resource utilisation (150 points);
- quality assurance of products and services (140 points);
- quality results (180 points);
- customer satisfaction (300 points).

This list shows that customer satisfaction is the most important indicator. These criteria and their detailed indicators have been used by companies to develop self-assessment tools. The 1993 winner of the European Quality Award, ICL, has used the criteria as a business model method. ICL is among the world-class manufacturers actively using awards as a means to evaluate and re-engineer their processes, see Figure 7.5. Their measurement system has integrated financial and non-financial measures and is based on the groups of results suggested in the model. The model is a complete summary of the entire process for business understanding. It needs to be considered in depth and will give managers at all levels insight into the business or operation in which they take part. Measures must be used that are strategically and customer driven.

The quality awards naturally focus on quality, but provide a useful contribution to business modeling. Many of the tools and methods originally developed for quality management could be leveled up to the corporate level and expanded with the other parts of the company.

Audit dimensions to ensure balanced business modeling

This methodology is developed as an answer to the aggravating business challenges companies are facing today. At the same time as costs are expected to decrease, value added for customers are expected to increase. New concepts like lean production and re-engineering of business processes are adopted by many companies. This is not synonymous with the traditional cost reduction strategies followed by many companies. The traditional strategies and modeling methods tend to focus mainly on cost to improve bottom line. However, this strategy has proven its insufficiency because of lacking a linkage between cost and income. We want to introduce a method that ensures focus on both parts influencing the bottom line: income and cost, that is improvement in the value chain.

7.4.3 Improvement in the value chain

The value chain is perhaps the business modeling method most referred to in management literature. It provides a useful step away from the organizational structure. However, the model is easily misused when linked to a departmentalized standard costing system. Misinterpretation of the term activities as functions will bring you back to the outdated organizational structure model.

Experience from development and analyzes in the Norwegian TOPP study in mechanical and electrotechnical industries shows that activities are easily interpreted as functions.

The value chain is primarily a strategic tool developed as a mean to identify sources for cost and differentiation criteria, and does not provide the final step into identification and management of business processes. Linked to more detailed methods it could be a valuable framework.

Porter emphasizes that each company has its unique value chain. His model is only the framework for definition of the company-specific value chain.

The value chain mainly emphasizes the day-to-day operation of delivering the product or service. Most important activities could be built into the concept, but it does not depict for instance the new product introduction cycle explicitly. This should not be understood as if it is forgotten by Porter, but as limitations in his visual model. Porter says that the emphasis of the value chain is to 'segregate a firm into strategically relevant activities in order to understand the behavior of cost and the existing potential sources of differentiation' (Porter, 1985). Lynch and Cross (1991) claim that the value chain does not adequately address the relationship between the specific activities. Again it could be stated that the simplicity of the concept calls for additional tools to ensure other aspects are included. The value chain is therefore a powerful framework and the required link to the strategic aspects of a company.

7.5 AN ORGANIZATION DESIGN MODEL FOR INFORMATION FLOW

Due to the continuous appearance of uncertainty in all processes, organizations have to make themselves adaptable to the ever-changing environment. The enterprise is referred to in terms of up and down the vertical organization in Galbraith's organization model. This environment includes customers, government, suppliers, market, technology, etc. As the organization's sub-tasks increase in uncertainty, more exceptions arise which must be referred upwards in the hierarchy. The hierarchy upwards will soon be overloaded if the amount of exceptions increases. This will create undesired situations that include serious delays, time to respond down the hierarchy, increase of throughput time in production, and more capital tied up. A dynamic organization has to be adjustable to the changes if they want to stay competitive.

7.5.1 The Galbraith organization design model

The Galbraith organization design model, is a model of five alternatives of how to organize the staff to create an organization to handle exceptional events occurring without overloading the upward referral system. Galbraith focuses on the handling of information as a very vigorous remedy for companies exposed to a rapidly changing environment.

The five alternatives for helping the company, presented by Galbraith (1977) are:

Alternative 1: Environmental management
Alternative 2: Creation of slack resources
Alternative 3: Creation of self-contained units
Alternative 4: Investment in the vertical information system
Alternative 5: Creation of lateral relations.

As we see from Figure 7.6, three strategies exist for reducing the need for information processing. The two latter alternatives represent strategies for increasing the capacity to process information. A short presentation of

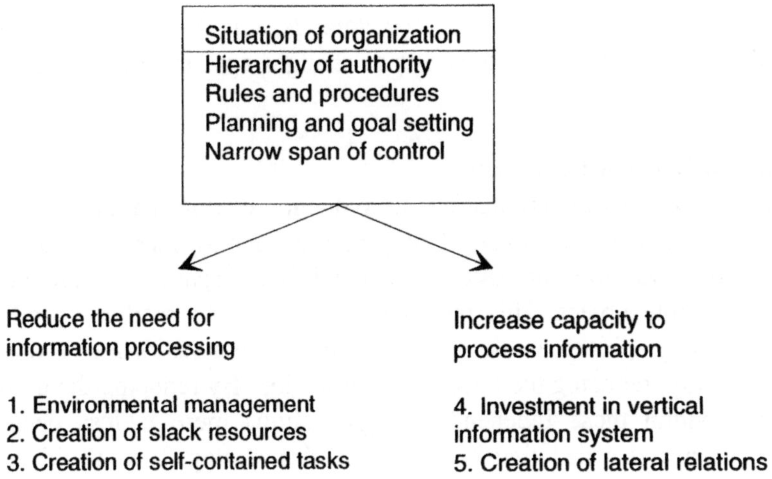

Figure 7.6 Galbraith's organization design model.

these follows. It is important to emphasize that this is not a choice for one of the alternatives, but there might also be a case of combination. For instance the cost involved by the implementation could be one of the factors deciding for which strategy to go for, resulting in a combination of two or more strategies to optimize the results.

Environmental management

Based on the evolvement of external business factors, the enterprise needs to have control over the crucial factors. This is possible for instance by creation of strong alliances with critical suppliers (vertical integration). If the enterprise can not adapt to the given market, strategic decisions can be made to concentrate on more adequate markets. Few enterprises have the capability to do any major changes in the market, and this is therefore not the appropriate strategy for most companies.

Creation of slack resources

Slack resources can be extra work hours, introducing high-capacity equipment, etc. By reducing the performance level, the number of unforeseen situations would be reduced. This is the risk of customer-driven companies, unable to adjust to all the variations in the market. Creation of stock of finished products is another possibility. Use of advanced systems for planning and production is another. Depending on how the management chooses to use slack time, it might result in decentralization or centralization of decision. For manufacturing companies the slack time might be used for absorption of uncertainties, but the capital tied up in inventories must not exceed the costs of changes appearing (specifications, etc.).

Creation of self-contained units

Within this strategy the emphasis is on reduction of information processing between units in the company. Companies producing many products, all influenced by variation of customer demands, can organize the production of the principle object. This means an orientation to markets, products, projects, geographical areas, etc. The small organizations can operate independently, reducing the need for coordination. By reducing the division of labor within these independent organizations there is less need for coordination and scheduling of work. By this strategy you avoid cross-functional specialization of the company's material and product flow. The point of decision is moved closer to the source of information to systematically reduce the information processing. This strategy has the drawback of generating extra costs because of necessary duplication of employees in the different units. It is also more expensive to buy several pieces of equipment than to buy one large piece of equipment of the same total capacity.

A reorganization of the entire staff might seem to the people involved, as a happening just arising when they have settled in their new positions. But constantly alteration of the organization creates a dynamic organization adjustable to changes arising. However, one has to be aware of possible loss of people not obtaining positions they felt entitled to have. On the other side, the staff will during these reorganizations gain a fresh overview of the company if they are placed in different divisions. A network of people knowing each other will also create lateral relations crossing the divisions' borders.

Investment in the vertical information system

New channels for communication, and the introduction of new decision mechanisms will increase the capacity of total channels for communication. The capacity of the organization will increase to make use of the information needed in the work operations. This is especially important for companies that want the flexibility to adjust to customer demands and promises this flexibility.

Galbraith (1977) describes four policy variables to bring the information, regarding the decision making, to the point of decision and to make the adequate decisions. These strategies consist of:

1. periodic vs. continuous information flows to and from the decision mechanism;
2. global vs. local data base available to the decision mechanism;
3. formal vs. informal information flows to and from the decision mechanism;
4. capacity of the decision mechanism.

The strategies point out the various alternatives for how to exchange information and the importance of availability of information for the decision makers. Critically it is also the capacity of the decision-mechanism to take the decisions when needed.

Creation of lateral relations

The process of going through channels to make the right decisions might seem bothersome, but is nevertheless necessary in most organizations. The use of this process could be substantially improved if the process were designed to fit the formal organization. Galbraith (1977), separates the different forms of lateral relations in order of increasing cost.

1. Direct contact between managers who share a problem.
2. Liaison roles to link two considerable dependent departments.
3. Temporary task forces to solve compound departmental problems.
4. Permanent teams for recurring interdepartmental problems.
5. Integrating role when there are problems with management of lateral processes.
6. A coordinating role to solve problems with substantial differentiation.
7. Dual authority relations can create the matrix design.

As you see, these relations can be realized with the right organization for all employees, and with rules for the information exchange between the different units in an enterprise. As Timmermans (1993) emphasises, it is not the volume of data that is the problem, but its complexity. He is concerned with information systems and here the complexity of the information is crucial for the design and simplicity of the system. If the objective is to create an automated information system then the complexity of the information is what is difficult to handle, and the Galbraith organization design model does not mention this.

7.5.2 The Galbraith organization design model in practice

Let us look at some examples to support the Galbraith organization design model.

Computer-aided manufacturing (CAM) is the term commonly used for all computer-controlled activities involved in actually transforming raw materials to finished products (Kochan, 1986). This includes computer numerical control systems (CNC), flexible manufacturing systems (FMS), robotics, computer-aided inspection, etc. A CAM system is limited to larger companies because of the high cost, but has shown increase both in productivity and quality. The main benefit we get out of CAM is increased labor productivity, reduced possibilities for transferring faults, healthier production, and so on. Less setup time is also an important aspect when atomization is introduced, and atomization creates more time for other activities.

The flexible manufacturing systems (FMS) aims at producing small batches of components, in a fully automatic mode on a single production line. Every part of the line is computer controlled and can be programmed to produce first one type of a component, followed by another without manual intervention. A central computer coordinates the operation of all the different elements and links up to the plant's main data-processing computer to provide reports (Kochan, 1986). As the name indicates these systems are designed to create flexibility for the user, and thereby better able to meet customer-defined products. FMS is though, only applicable for certain manufacturing systems.

Integration of systems filling the same function during the process of developing new products, is characterized as horizontal systems, e.g., computer-aided design (CAD) systems. Computer-integrated manufacturing (CIM) constitutes both a horizontal and vertical integration. This integration is to optimize the process of developing and producing new products, by integrating the necessary functions. The CIM system has to be developed for the specific company, by coordinating the company's systems as in CAD/CAM, robotics, FMS, FEM (finite element method), etc. A series of applications connects all these modules together in a CIM system, and the most important ones are (Bø, 1989):

- product models and data bases;
- standards for networking and interfacing;
- user communication;

- administration systems;
- functional systems (e.g., accounting).

CIM allows the accountants, operators, production planners, and engineers to use the same data base. To succeed within the CIM concept, the integration of the entire organization and all the workers is essential. CIM becomes one of the important factors for breaking down barriers between functions/departments.

In a CIM environment, computer-aided design (CAD) technology is the ability to create automatically a product data base consisting of design data, dimensions and key characteristics, as well as associations with other elements of a product (Kochan, 1986). One of the approaches for using CAD systems is to interface different CAD systems so that for instance vendors are able to exchange information. The ISO/STEP is working heavily in this area.

A CIM system will consist of a computer hierarchy. For the management information systems, it is assumed the existence of computers at three different levels:

- central computer facilities servicing the whole company;
- local computer facilities available for performing tasks within one department;
- cell computer facilities serving as master control unit for a cell installation.

It is often difficult to integrate production planning and control in a CIM system, because of little standardization in representing shop floor control data. If this can be realized, integration of production planning and control in a total CIM environment would be very valuable if it is integrated with the existing types of systems.

7.6 AN OVERALL MODEL FOR MANUFACTURERS

Trying to link different parts of the company can be a very challenging job. Here we show a model that has the objective to take the product data and information through the process of product definition.

7.6.1 An integrated data model

The primary purpose of an integrated data model is to provide a generic, reusable representation, also known as an ontology, that integrates the representation of product, process and organization knowledge. This includes (IMS Globeman 21, 1994):

Product:

- features
- versions/revisions
- requirements, goals, constraints
- documentation
- design rationale
- function, behavior

Process:

- activities
- resources
- goals, constraints, milestones

Organization:

- structure
- resources
- goals, constraints
- responsibility
- authority

This PPO model illustrates the process of product development, from idea formulation to specification of the final product, constrained to the customer's needs (Christiansen, 1993), see Figure 7.7. Customer needs can for instance be found through QFD (Quality Function Deployment, House of Quality). The next step is to take into account the processes that are to be used for the manufacturing of the product. Which activities are required and how can we control them to ensure quality of the product? The third part of this simplification of the product development process, is to incorporate the organization. The employees and the relationships among

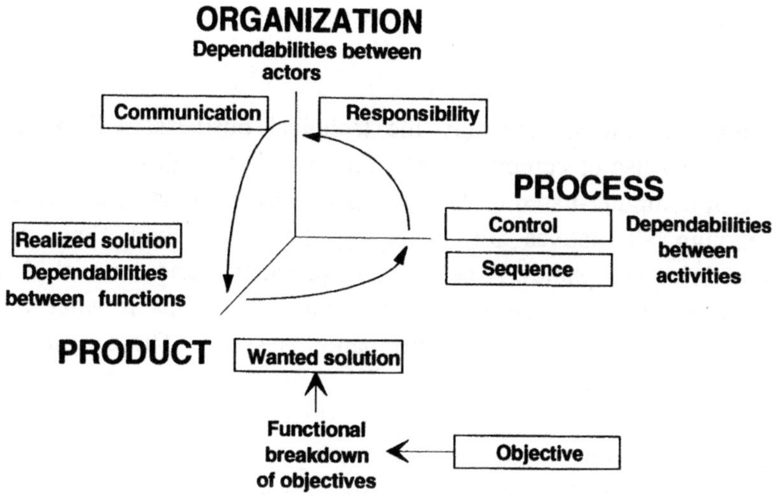

Figure 7.7 A PPO model.

them are vital for manufacturing performance. Communication is the visible form of information flow in an organization. Through TQM and focus on countinuous improvement, we have learned that the responsibility of each worker is essential for success. The organization of employees, and how they create lateral relationships are necessary to take into consideration. This implies that how the employees are organized is determinant for the performance both for the organization as well as for the individual employee. When going through all steps in this model, maybe several times, you end up with the final product that is put into production. This type of cycle in the product development will take into consideration all the functions that have an impact on the product. This kind of model can be used for creation of a virtual product. When going through this cycle ahead of production startup, you can prevent surprises when it comes to production, or in the worst case when the product enters the market.

The PPO model is quite a simple representation of the company and its main entities. For each axis the purpose can be described, and even changed to fit new products.

Teamwork is relational; the quality of performance rests on the quality of interactions, communication, and coordination among team members. For larger companies or for companies working in a network of supplier/customer relations the need for tools to handle this situation is obvious.

7.7 MODELING FOR FUTURE MANUFACTURING COMPANIES

The future business, especially the technical business, will be greatly influenced by use of computer-tools. The goal is though not to implement the use of computer-based tools in all operations, for today we have simply not enough alternatives for managing the information flow through manufacturing companies. A new type of production and material planning and control system is required in the near future. This is what we call fourth-generation management information systems. These systems are based on recognition of:

- new production principles (no stock, group technology, etc.);
- changed information requirement (improved communication between man/machine, consequent analysis in planning);
- new system design criteria (decentralised systems, user-developed systems, prototyping);
- new tools (computer graphics, text processing, high level languages).

Here we present the current practises and prospects for leading manufacturers in information management systems. To be flexible and useful for production companies they require development of comprehensive models from several manufacturing aspects including simulation and virtual manufacturing.

7.7.1 Simulation

In the manufacturing business, especially in the recent years, simulation has become a part of everyday life. Both for short-term and long-term planning, we can simulate all activities in the manufacturing process to reach for the optimal use of resources. For large, complex production plants, a simulation tool is absolutely needed to control the total process. Objectives or purpose to use simulation can be split into:

- control of manufacturing processes;
- planning a manufacturing system.

Simulation acts as decision-support to production for, according to Rolstadås (1986):

- schedules
- change in capacity
- use of overtime
- priority rules or sequencing
- delivery times
- stock levels
- use of subcontractors
- labor allocation
- use of alternate routings.

Simulation for system design is to optimize production systems (e.g., production layout, machine capacity).

We know that electronic simulation reduces time, cost and risk, while providing information needed for continuous product improvement. In simultaneous distributed work, simulation is not a stage in the development process, as was physical prototyping; now simulation becomes a continuous aspect of shared, continuous design. Although simulation tools are becoming very popular in manufacturing systems modeling, there is a drawback with such tools because they require extensive programming for building models describing the manufacturing system. The programming needs expertise both for the actual manufacturing system, and a programming language to get an accurate simulation model. This complex task is only recommended for large batch manufacturing or manufacturing that is too complex to handle manually. Partial simulation of for instance one production line can easily be implemented for *ad hoc* simulation.

Simulation of a manufacturing process or other related activities are only adequate when realistic data is used, and in most cases this is real-time input. Here is where the problems have emerged up till now. Automated data acquisition is, unfortunately, not a practically good solution for most manufacturing enterprises. It is a very cost and time demanding process, as well as complex. If the input data is not accurate, the output data must be regarded as suspect as a result of this and can not be 100 % correct as a solution. Compensating for the varying quality of the data, simulation results can be a very helpful tool in planning and in control.

For the future manufacturer we see simulation as the only method for planning and scheduling of the production process. There will be a need for large machines to handle all information on products and processes previous to the startup of making new products. This is the only possibility to get the best use out of the resources available. This is what we call virtual manufacturing. All the objects and resources related with manufacturing should be completely modeled in a computer. Every necessary manufacturing task should be simulated as precisely as possible, and target products and associated manufacturing data can then be virtually produced and verified before making real products.

Simulation generally needs models of what is going to be simulated. To simulate a CIM system there exists several techniques to build appropriate models for the requirements.

Structured Analysis and Design Technique (SADT) is a total technique for design and analysis of a system. A SADT model is a description of a system, and provides help through:

- systematic to solve problems;
- a language that expresses complex coherence and suitable for human communication;
- work routines to achieve high quality results and clear responsibility sharing and management.

IDEF 0 is a method which is used to produce a structured, graphical representation of the process concerned with a manufacturing system or environment and of the information and objects that interrelate those processes. IDEF 0 is a subset of SADT with the purpose of offering structured methods for efficiently developing information systems for manufacturing environments (Myklebust, 1992).

GRAI is a system development technique developed by the University of Bourdeaux, France, especially for production management. Here, system development techniques can be characterised by:

- decomposition principle;
- graphic tools;
- conceptual models.

The method generates user viewpoints to be included in design specifications.

FOF Production Theory was an ESPRIT research action concerned with one-of-a-kind production and products in the Factory of the Future (FOF).

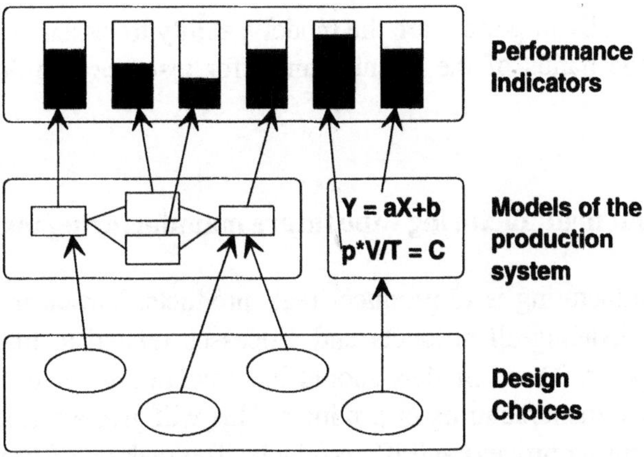

Figure 7.8 Components of a design reference model.

As a part of the work a model of a manufacturing company dealing with one-of-a-kind production was developed. The model formed a baseline for a consistent theory, as well as designing, redesigning or testing a manufacturing system. The FOF project was specifically looking at a computer-integrated manufacturing (CIM) environment (Rolstadås, 1991). The model includes decision-support based on:

- product changes;
- resource changes;
- production layout;
- resource organization;
- decision rules and criteria;
- management systems;
- etc.

For a one-off production system, simulation of effects when changing design choices is a powerful tool. The idea is that when design choices are changed, this will influence the end result (Rolstadås, 1991). This process is split into two. First the design changes will induce changes in the production system, and this will change the performance indicators at the user, see Figure 7.8.

The model of a production system is built to connect the various influencing factors of the production system. This will give the effects of the change in the design phase a different result than before. The quality of the changes in performance indicators is dependent on the quality of the model. Especially important are the model's ability to lessen the effects that are out of the hands of the manufacturer, for instance late delivery from vendors.

7.7.2 Virtual manufacturing - the future manufacturing management?

Virtual manufacturing is to produce 'real' products. Virtual manufacturing starts with modeling all products and processes related to manufacturing, into a computer. These models should be simulated to give a very exact picture for the manufacturing of products. This will give a virtual picture of the manufacturing process and it's products. The making of prototypes will no longer be necessary and planning of the manufacturing process is directly put forward through the virtual modeling. When you make the real products you correlate the virtual model with the real world to provide the accurate information. This will connect all the stages involved in the process of making products to use a computer as a tool for performing every manufacturing action. Future manufacturing like virtual manufacturing, makes the pre-production demand for information and data even bigger. Some parts of the virtual manufacturing concept have been developed, but it is still a far way to go to reach the finite goal of this new 'technology.'

The core of the virtual manufacturing consists of modeling software. The three major models discussed in the two sections above, product, process and activity, have the ability to define, describe, generate, visualize and operate manufacturing entities in computers. We have to combine these modeling programs to generate an entire manufacturing process in VME. For instance start a particular process by using the data/knowledge base of a particular manufacturing environment, and transfer the virtual manufacturing process to the real manufacturing process. Therefore a realisation technology for the virtually modeled world is required but is not yet fully developed.

The main question here is how to organize intelligent, complex, individual machines and information-processing equipment into a manufacturing system. The establishments of organization methods that eliminate solitary islands of automation, assure high ease of construction of manufacturing systems and provide sophisticated tools for globalization will be required.

We require technological approaches for realizing decentralised manufacturing activities to cope with globalization (Prytz, 1993). The communication channels should establish skilful linkages between distributed activities across countries. Geographically-dispersed activity demands innovation in the functions of information management, planning, logistics, R&D, design, etc. They should be supported by modeling methods for generating and modifying a manufacturing organization to meet individual conditions. Thus companies will be guaranteed to make plans of their global infrastructure to minimise structural costs and accommodate their organization and operations to new environments.

Diverging consumer needs make it difficult to keep traditional ways of production of limited kinds of products. High levels of customer satisfaction can be obtained by customer-oriented design. We require innovative manufacturing systems which can adapt themselves with various types with quick response. Concurrent engineering is regarded as quite an important technological concept to solve the problem and is commonly focused on by all the manufacturing domains. Also configuration management tools/methodologies will be developed, which allow us to reconstruct manufacturing systems easily to meet customers' high level requirements.

The future ideal company is the kind of paper-free environment, where all communications run through networks and other telecommunication possibilities, for instance teleconferences and photo-phones. The cost of producing paper like present business do, will no longer be preferable based on environmental and time pressure issues.

An important aspect is that increased communication through computers and telecommunication networks must not break down human relations. This is necessary for most workers as a strong motivation factor. Enterprises connected in world-wide alliances and working inter-enterprises, will be the prototype for this development. The development of these technologies is supported by new modeling approaches for manufacturing aspects. Helpful tools that can help humans to control and handle complex situations are required. The nearest future will show the results.

8. PERFORMANCE MEASUREMENT

Nobody argues against the importance of measurement. However, measurement is a disputed and misused term. Some authors like Sink and Tuttle (1989) claim that you cannot manage what you cannot measure. Others may disagree with this philosophy due to the danger of choosing the wrong metrics. Development of an effective measurement system is a crucial task for every company exposed to tough competition. A company must know were and how to improve in order to have enough resources to secure long-term survival.

A major challenge for every company is to develop a representative measurement system. Development is probably only possible trough a trial-and-fail process. Design of a complete measurement system on the drawing board is utopia considering the complexity of a modern company.

8.1 TRADITIONAL PERFORMANCE MEASUREMENT

Every company has a measurement system of some kind. Companies are required to report financial key figures to authorities, shareholders, etc. An impressive amount of financial metrics and figures are available in most companies. The purpose is to provide a correct picture of the financial performance. These figures are often viewed as synonymous with performance of a company. This was true in a stable industrial environment, but could be questioned in a period with rapid change.

Existing measurement systems are often dominated by financial metrics easy available from accounting systems. Measurement is often limited to collection of results, which is a retrospective approach. Choice of metrics is often closely related to the traditional planning hierarchy of a company. A typical model with major information flows is shown in Figure 8.1. The traditional planning hierarchy is characterized by a decomposition of objectives as plans with related aggregation of measures as feedback. Decomposition and aggregation are based on an organization model divided in departments.

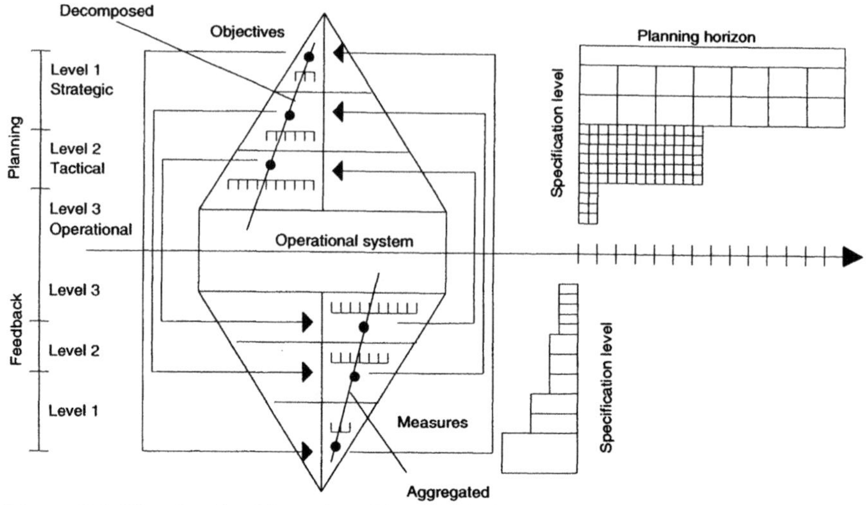

Figure 8.1 The planning hierarchy with major information flows.

The ability to aggregate has been the best argument for keeping focus on financial indicators in performance measurement. Most operational indicators for flexibility, quality and time could be difficult to aggregate and often play a minor role in performance evaluation. Aggregation of departmental performance on the corporate level was a driving force in development of standard costing. Introduction of a process view makes this approach obsolescent. Performance improvement in interdepartmental processes requires interdepartmental performance indicators.

Paradigms about measurement

Measurement is disputed in many companies. These disputes are often caused by differences in paradigms. Traditional paradigms about measurement could be serious roadblocks or obstacles to effective performance measurement and improvement (Sink and Tuttle, 1989).

- **Measurement is threatening!** Measurement has traditionally focused on controlling employees. The point to be made is that measurement is not the object to blame. A clear distinction has to made between measurement and evaluation. The evaluation process has to be credible.
- **Precision is essential to useful measurement!** This might be true in technical or even accounting situations, but not for performance measurement. The object of performance measurement is to tell us where and whether we are improving and not the exact level. Putting too much

effort in developing precise measures can actually prevent the introduction of a balanced measurement system.

- **Single indicator focus!** There is a tendency to seek out a single, bottom-line indicator attempting to tell the whole story about the measured matter. Performance is a complex matter and a set of balanced measures are required in order to give a meaningful picture.
- **Overemphasis on labor productivity!** For many products, labor costs account only for a small part of the total costs. Still, labor productivity is getting disproportionate attention in performance-improvement work. We have to shift from a single-factor paradigm to a multi-factor paradigm, where other, often far more cost-contributing factors, are given the appropriate emphasis.
- **Subjective measures are sloppy!** Our 'bean-counter' tradition of looking at money as the only reliable measure has prevented softer dimensions in measurement systems. However, good results are obtained in the area of psychology concerning soft dimensions. Refusing to incorporate soft dimensions in the measurement systems prevents development of a balanced system.
- **Standards act as ceilings on performance!** People tend to look at standards as the upper limit for performance. Fixed standards imply that once the standard is met, there is no need for more effort in further improvement. This is often seen when incentives are related to standards, because boosting the standard means renegotiation of the standards. A reference for performance evaluation is required, but uncritically linking measurement and evaluation in standards is not a credible approach. Other factors may influence the performance and evaluation is required.

8.2 ELEMENTS OF A PERFORMANCE MEASUREMENT SYSTEM

8.2.1 The framework for a performance measurement system

Purpose of measurement

As a basis it could be stated that measurement is a non-value-adding activity. However, if a specific measurement is required by a stakeholder it certainly is a part of the value adding process. The reason for this discussion is to ensure that development of a measurement system is linked directly to competitiveness and improvement. Cost/benefit has to be carefully evaluated when developing metrics to ensure a lean measurement system. Well-defined purposes makes it easier to:

- **Choose the right metrics**. A huge number of metrics are available. A well-defined purpose helps to choose the best.
- **Prioritize indicators**. Measurement is expensive and indicators have to be prioritized to identify the most important ones.
- **Choose appropriate measurement method**. Choice of measurement method is related to the purpose.
- **Achieve understanding and acceptance.** Without acceptance by people involved the quality of the measurement could be unpredictable. It is impossible to develop a measurement system that cannot be manipulated.

Performance measurement often has a bad reputation both at individual and organizational level. It has traditionally been related to performance-based salaries and control of blue collar worker with the old productivity paradigm. Important purposes for a performance measurement system are:

- **Decision support**. Ideally measurement should indicate where to act, perhaps how to act and hopefully monitor the effect of the action. Decisions should be based on knowledge and measurement plays an important role to provide information.
- **Monitor effect of strategic plans.** Implementation of strategic plans has to be monitored to make necessary corrections to ensure achievement of long-term goals. Indicators have to be chosen to monitor consequences and achievements.
- **Performance evaluation**. Evaluation is required for a number of reasons as tracking improvement potentials, setting new yardsticks, satisfying requirements from stakeholders, distributing incentives, etc.
- **Diagnosis.** A company needs indicators with a diagnostic purpose. If business achievements are decreasing, the performance measurement systems should be able to give some warnings in advance and provide input to a search for reasons. However, it is difficult to isolate the cause/effect relationship.
- **Manage a continuous improvement process**. A continuous improvement process often provides stepwise 'blue savings' like released capacity, reduced future costs or increased value for the customer. Measurement is important to justify further investments and

effort in the process, to manage the process and ensure consistency with strategy and to transform improvements into business achievements.
- **Motivation**. Measuring progress is necessary to justify further effort in the improvement process. Resistance against change is considerable in most organizations and lacking progress is enough ammunition to kill a project or a process.
- **Comparison.** Evaluation of performance and performance planning depends on a reference to identify performance gaps. Comparative benchmarking is possible to identify these gaps.
- **Record development**. Documentation of development could be demanded by stakeholders like customers, authorities, alliance partners, etc or used actively as a marketing tool. Measurement to monitor and record supplier performance to give input to their improvement processes could result in productivity improvement.

Taken into account this extensive list of purposes it becomes evident that developing a measurement system is a comprehensive and difficult task. Our list with purposes is only meant as a reminder of the great variety of considerations in order to develop an effective measurement system.

Definition of a performance measurement system

Every company has a unique structure consisting of facilities, equipment, products, competence, and infrastructure. Activities and processes could be universal, but the structure defines the identity of a company. A company is surrounded by an environment that is continuously influencing the development of the structure.

Performance in a business context is defined as the efficiency, effectiveness and adaptability of a company. Business performance depends heavily on the match between the structure and requirements from the environment. All elements of the structure should fit perfectly to each other as illustrated by the jigsaw in Figure 8.2. The infrastructure is the heart of the structure and is responsible for the integration of all the other parts to match requirements from the environment.

Effectiveness is determined by the extent to which requirements from the environment are met, whereas efficiency is defined as how economically the resources of the company are utilized. Effectiveness depends on the match between the structure and the environment, whereas efficiency depends on the internal match between the elements. The third dimension covers the adaptability to changes in requirements from the

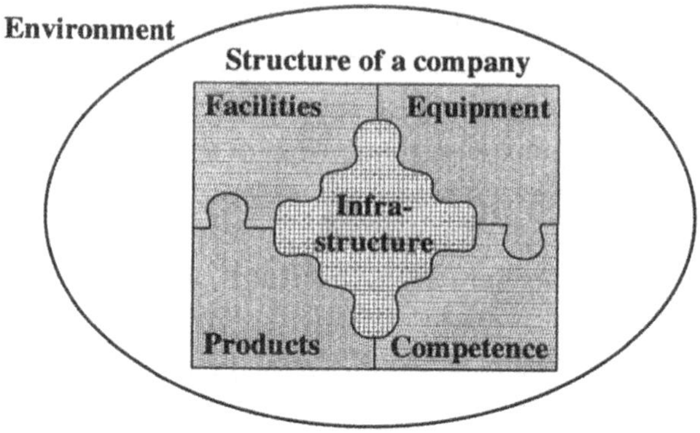

Figure 8.2 The structure of a company.

environment. It is absolutely necessary to ensure that the flexibility is prioritized in structural decisions. The infrastructure has a key role in the adaptability due to its integrative role.

Adaptability makes no sense at the operational level and performance is defined as the efficiency and effectiveness of actions. Performance measurement is defined as the process of quantifying actions. However, characteristics of actions on a strategic level differ dramatically from operational actions. Performance measurement will then be quite different for the different actions. It becomes evident that we are not talking about one measurement system but the presence of several layers of systems as shown in Figure 8.3.

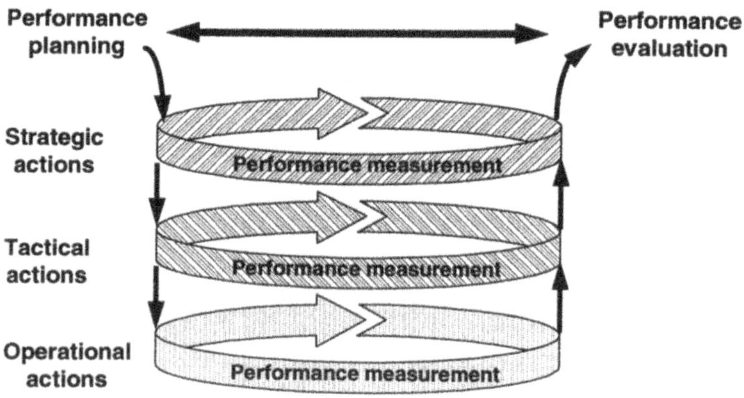

Figure 8.3 An integrated layer of performance measurement systems

Figure 8.3 shows an integrated layer of performance measurement systems. This model differs from the traditional planning hierarchy in the sense that perfect decomposition of plans and aggregation of feedback is not a requirement. Aggregation is done where it makes sense, because this is an efficient way to handle information. However, we will emphasize that measurement and evaluation are independent activities. Evaluation could be based on input from different layers of 'measurement systems' depending on the purpose. Approaches like sampling or more subjective aggregation could be used to serve purposes with evaluation. Performance measurement should be faithful to purposes previously adopted.

A performance measurement system is a set of metrics to quantify actions. Figure 8.4 shows a simplified framework for a performance measurement system. The performance measurement system should match the requirements from the environment. It should reinforce the strategies, match performance priorities, match the culture, match existing reward systems, match requirements for documentation, etc.

A performance measurement system has a hierarchy of metrics as seen in Figure 8.5. We distinguish between external indicators and internal indicators. External indicators are supposed to support the external performance priorities, whereas internal indicators are supporting the external. Mathematical relations between the different levels are not required. A performance measurement system is divided in two parts:

- **Stable part.** The stable part is a 'fixed' set of metrics which is aimed for performance evaluation and different kinds of reports
- **Temporary part.** A set of preliminary metrics to monitor and support specific problem solving or improvement projects.

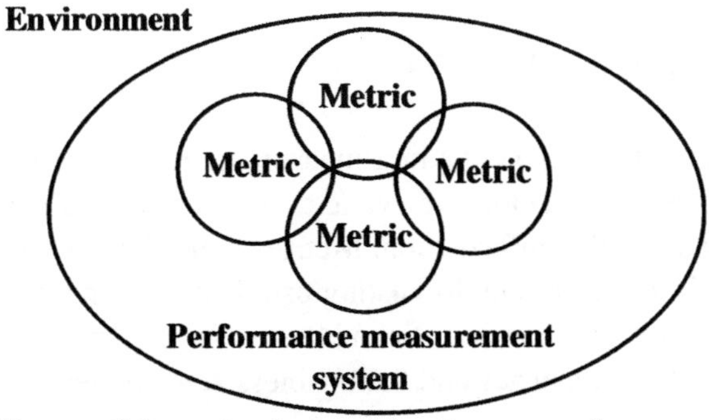

Figure 8.4 A framework for a performance measurement system.

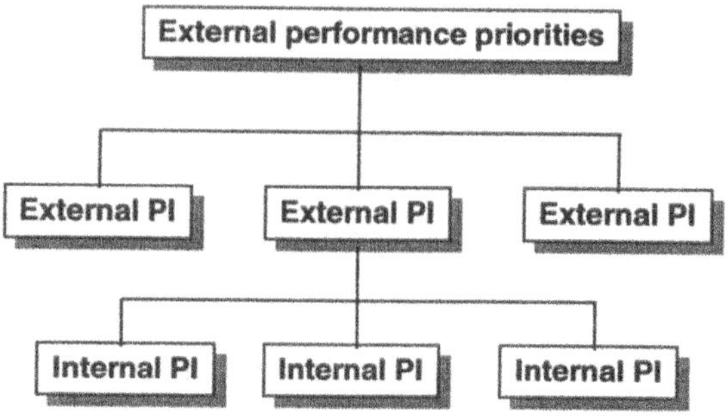

Figure 8.5 Simplified hierarchy in a measurement system.

We have made a dangerous simplification to visualize the difference between the two major purposes. An efficient measurement system has to be maintained continuously to sustain capability in a rapidly changing environment. Reports for owners, top management, customers, suppliers etc., are improved as the company gains more knowledge of their need of information. Improving this knowledge and the communication to the stakeholders is a critical task in marketing the company.

The preliminary part of the measurement system is problem-oriented and is continuously tailored to follow up actual problem solving or focus areas. Measurement is then actively used to manage and control improvement. Measurement could be performed both in advance, during projects and afterwards to ensure successful implementation. A company has limited resources to use for measurement and areas for measurement have to be carefully chosen. When a problem is solved the control measures are usually removed. If one of the temporary measures increases the knowledge of total performance it could be adapted in the 'stable' part.

8.2.2 Different types of metrics

Metrics are the building blocks in a measurement system. We use the term indicator for an information we want to know. A metric is the specific method to quantify this information. Metrics could be classified in different ways. First, it is important to distinguish between the various physical characteristics of the metrics. Secondly, it is important to explain the connections between metrics and the business achievement.

Table 8.1 Differences between hard and soft metrics.

Hard metrics	*Soft metrics*
Objective reference	Observer bias
Accurately known	Surrogate indicator
Hierarchical	Multi variable situation

'Hard' versus 'soft' metrics

Hard metrics are pure facts that are possible to measure directly whereas soft metrics are intangible metrics that have to be measured indirectly like for instance attitudes. They have some very different characteristics as seen in Table 8.1.

Hard metrics are the far most widespread metrics. Soft metrics and intangible metrics are considered to be sloppy by many practitioners. They rely only on tangible numbers with an objective reference. However, as Deming (1986) emphasized, the most important numbers are often unknown. Management by numbers is one of the deadly diseases that have ruined several companies in the Western world. Customer satisfaction could be an example of the insufficiency of hard metrics. Simply measuring the number of complaints and assuming that those not complaining are satisfied is at best naive.

Both hard and soft metrics have their strengths and weaknesses and should be evaluated according to purpose of measurement. The most important question is what knowledge measurement should provide.

Financial versus non-financial metrics

Financial and non-financial metrics are a subset of 'hard' metrics. According to the old paradigm, money is synonymous to performance. Most indicators are somehow translated into financial terms, otherwise they are not taken into account in performance evaluation. Performance, when defined by metrics about output, can only be retrospective. Integration of financial and non-financial metrics is necessary to execute real performance management. Table 8.2 provides some examples of traditional financial and non-financial metrics.

Most metrics within the table are retrospective and internally focused. It could be stated that we have chosen these deliberately, but several case studies have confirmed this impression.

Achievement versus process metrics

This confrontation could be stated to be one of the main difference between Western and Japanese measurement systems. Western management culture

Table 8.2 Examples of traditional financial and non-financial performance metrics.

Financial performance metrics	*Non-financial performance metrics*
Budget vs. actual variances	Inventory turnover
Product/product group profitability	Labor efficiency
Cash flow	Capacity utilization
Return on total capital	Defect ratio
Overhead absorption	Lead times
Customer profitability	Delivery precision

emphasizes achievements, whereas the Japanese include process metrics in their evaluation of performance. A process measure is intended to describe important characteristics of the process that is believed to have impact on output. An example could be number of meetings in cross-functional groups. The Western focus is to measure number of implemented changes.

In Western companies, performance is often synonymous with financial efficiency and hence achievements. Results are expected to come immediately and projects have a short payback period.

Japanese companies focus a balanced scoreboard that includes process metrics. A good process is believed to give results eventually. A long time horizon is accepted and all kinds of improvements are continuously sought.

8.2.3 Develop a consistent measurement model based on the three performance dimensions

We have developed a measurement model based on the three dimensions of performance; efficiency, effectiveness and adaptability. In this way all dimensions of business achievement are covered:

- achievement metrics - direct metrics for business achievement;
- diagnostic metrics - indirect metrics for business achievement;
- competence metrics - capability of future business achievement.

Achievement metrics

Achievement metrics are direct metrics for actual business achievement. They are hard facts and could be measured directly without any interpretation. Financial metrics dominate and typical examples are:

- net profit
- return on investment

- market share
- export share.

Many claims that these metrics are the only valid metrics for performance since the bottomline has to be the main focus for every company. However, these are lagging metrics and they are heavily influenced by external factors as changes in currency, customs duties and political factors. At best they can say something about the match between requirements and performance in the past period, but little about the match for the moment and certainly not for the future. Financial achievement is not synonymous with performance, but good performance is necessary to achieve financial success. A major problem with pure financial metrics is that they are often drivers for short-term optimization.

Achievement metrics are collection results and give no explanation of how to react on a negative trend. Although they identify the need to react they hardly represent an early warning. A decrease in customer satisfaction is not necessarily visible in figures defining market share or net profit before great damage is done. Achievements are necessary for survival, but represent only a part of a measurement system.

Diagnostic metrics

Diagnostic metrics are indirect metrics for business achievement. They are critical success factors for competitiveness without a direct link to traditional financial achievements. Critical success factors are defined as:

> *Critical success factors are those characteristics, conditions, or variables that have a direct impact on your customer's satisfaction, and hence your own success.*

A diagnostic measure should be able to explain trends in achievements and perhaps suggest suitable actions. They are a vital part of the early warning system together with for instance market surveys and competitor analysis. Typical examples of diagnostic metrics are:

- delivery precision;
- delivery flexibility;
- product quality;
- product reliability;
- lead times on customer requests;

- customer satisfaction;
- outstanding claims.

Most diagnostic metrics are non-financial. However, they may have great impact on financial achievements. They focus on the ability to meet stakeholders' requirements. Customers have an exceptional position among the stakeholders and nobody should be allowed to forget that the payment ultimately comes from the end user who actually buys the product. Integrating stakeholders' requirements in the measurement system ensure diagnostic indicators directly linked to business achievement.

Another important aspect with application of external diagnostic metrics is the marketing effect. To increase income a company needs to communicate the improvement to the market. It is not likely that the customer will discover improvements unless he has a supplier evaluation system. Recording improvements of critical success factors is important both to monitor the development and to get credit for them in the market. Diagnostic metrics have an offensive role in order to identify improvement areas to exceed requirements from customers and outrun competitors.

Competence metrics

Competence metrics are the most difficult to define. They should describe how well the company is prepared for the future or to meet new requirements. Nobody is fortunate enough to know what the future will bring. Uncertainty is a natural part of the competence metrics and is probably the reason why such metrics seldom are applied in industry. Typical metrics could be:

- investment level in product development;
- time to market with new products;
- stage in the life cycle of the product portfolio;
- investment in manufacturing technology development;
- attitude towards change;
- flexibility to manufacture completely new products;
- percentage of work force in teams;
- level of training.

There exists no correct level for these metrics. Market surveys would not be of much help because customers probably have no opinion regarding this. It is up to the company to define the right level in performance planning. However, some of these metrics are of interest for an organizational buyer considering long-term relationship with the company. Competence metrics show how well prepared the company is to stay competitive. Recording and communicating the level of competence metrics is an important part of the marketing to persuade customers to commit for a long-term relationship.

Definition of the 'right' level of these metrics has to rely on intuition and not on standards. Aspects like product complexity, competition, life cycle time, innovation time, organizational culture, strategic objectives etc have to be considered. Benchmarking and other comparative tools and methods could be used to define objectives.

Relationship between result, diagnostic and competence metrics

A measurement system should contain a balanced mix of achievement, diagnostic and competence metrics. They serve different purposes and have different time horizons. Their validity as performance indicators change over time as simplified in Figure 8.6.

Achievements have limited predictive validity for future performance. Validity is decreasing as we approach the time for the measurement due to inertia of consequences with decreasing performance. Customers are committed by contracts and consequences are delayed.

Diagnostic metrics are most valid at the time for measurement. Validity will decrease over time due to changing requirements. Competence metrics are strategic because of the long-term horizon of actions and resources required to change adaptability. Efforts to improve competence metrics often have a negative impact of achievements in the short run due to

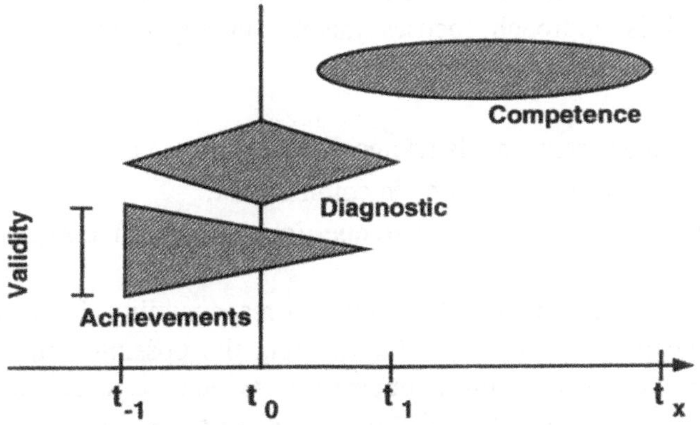

Figure 8.6 Validity horizon of the different metrics.

investments. Competence metrics are most valid in the medium term. Determination of medium term varies from company to company depending on their general planning horizon. Electronic industry and air craft industry are two opposite poles in such matters. Uncertainty about future requirements is the reason for the decreasing validity.

Simplified, it could be stated that the competence metrics describe the long-term means to achieve the wanted level of critical success factors that consequently give the required achievements.

8.3 DEVELOPMENT OF A PERFORMANCE MEASUREMENT SYSTEM

8.3.1 Different development approaches

There are three different approaches to develop a consistent performance measurement system

- top-down cascading method;
- bottom-up process;
- combination of bottom-up and top-down.

The top-down cascading approach is perhaps the most widespread. Management by objectives (MBO) is based on this approach. MBO starts with a definition of desired output followed by a cascading process through the formal organizational structure. The formal manager/subordinate relationship is the main transfer link. Budgets often drive the objectives and the required improvement. Feedback follows the same route in the opposite direction. This approach focuses mainly on departmental performance. Necessary success factors are:

- articulate the system in a language that all staff can understand;
- explain (not prescribe), to obtain commitment;
- mandatory right for each level to question validity of the explanation.

A bottom-up process is based on personal responsibility. Every member of the organization is responsible for defining the possible improvement in their area of responsibility. Improvements are accumulated and the corporate goals are determined. Improvement drives the budgets, because improvements and actions are defined in the process. The feedback system is a consequence of the objective hierarchy, but is not specifically discussed in the process.

The combination approach tries to unify the best from both, which isolated have several major weaknesses. There are several combination opportunities. One opportunity is to perform both approaches and identify the best combination in a joint process involving for instance a representative committee. Another opportunity is to design a process where the different people within the organization have specific roles.

We have applied the combination approach in performance planning to define performance priorities. These priorities are used as input in development of the performance measurement system. Performance planning and development of a performance measurement system are integrated and strict borders are not desirable.

The combination method integrates the process view into the objective and feedback hierarchy. Application of a strict top-down approach could be counter-productive in the sense that it fails to support interdepartmental improvements. A more horizontal focus is required to support this. An example could be a purchasing department measured by price variances and not the total purchasing cost including storing, quality, delays etc. The following quotation from Lynch and Cross (1991) could illustrate some of the difficulties:

> *Metrics are often fragmented and isolated by functional specialty. For instance, quality metrics are measured and reported by the quality department one week; cost metrics are done by the financial department and reported another week; delivery metrics are done by operations and reported daily. The result is that many of today's performance metrics are promoting and rewarding behavior that hampers a company's overall performance.*

Involvement of many people in the bottom-up process could be beneficial in several ways:

- exploit the knowledge of the whole organization;
- generate ideas for improvement;
- reveal potential areas for improvement;
- ensure a comprehensive view for improvement;

- achieve commitment and participation that promotes cultural development;
- develop a measurement system that people trust and fulfill the purpose;
- develop a more efficient information flow.

However, a process initiated from the top management is necessary to identify the main performance requirements and ensure commitment. The challenge is to unify both approaches into an effective measurement system.

8.3.2 Requirements for a performance measurement system

Every company has a measurement system of some kind. Development of a performance measurement system is then a transformation process based on the existing system. What we want to advocate is to be more conscious about the system and the evolvement process.

There are several pitfalls in the process. Lawlor (1985) has identified some concerning measurement systems:

- The measurement system must be used and lead to action. Measurement for its own sake should be abandoned.
- Information that is produced only when someone asks for it is generally ineffective.
- Productivity metrics must be believed by and understood by those who have to use them.
- Too much information and too little information are equally bad. The former creates confusion and the latter blind ignorance.
- Aggregating figures to give averages may obscure significant variations in performance.
- Many productivity indicators are only partial metrics and therefore reveal only part of the story.
- The measurement system must not be believed to be perfect, and left to remain that way for ever.

There exist several approaches to development of a measurement system and hence guiding principles. Some common wisdom is important to take into consideration:

- Measure what is important and not just what is easy to measure.
- What we measure is perceived as being important - therefore we must measure dimensions that impact the attainment of our goals.
- Do not attempt to achieve a single measure, but use multiple indicators
- Look for behavioral consequences; potentially disfunctional consequences of performance measurement should be avoided
- The process of measurement must be clearly separated from the process of evaluation.

Transformation of the measurement system could be described by change in the following characteristics in Table 8.3 (Lynch and Cross, 1991). Our discussion so far could be summarized in the following requirements for a measurement system:

- integrate stakeholders' requirements
- support strategies
- reference for performance evaluation
- reinforce improvement
- consistent
- record development
- cost-effective.

Table 8.3 Transformation from traditional to a performance measurement system.

Traditional	*Performance*
Financial focus	Performance focus
Financially driven (past focus)	Stakeholder-driven (future focus)
Limited flexibility	Flexible, dedicated system for operational control
Not linked to operations strategy	Tracks concurrent strategies
Used to adjust financial	Catalyst for process improvement
Locally optimized	Systematically optimized
Decrease costs	Improve performance
Vertical reporting	Horizontal reporting
Fragmented	Integrated
Cost, output, and quality viewed in isolation	Quality, delivery, time. and cost evaluated simultaneously
Trade-offs unknown	Trade-offs addressed
Individual incentives	Group incentives
Individual learning	Organizational learning

8.3.3 Transforming the existing measurement system into a performance measurement system

We have developed an integrated approach for development of performance planning and performance measurement. A common understanding of performance priorities is the input from performance planning. These priorities have to be matched by indicators in the measurement system. However, the development process has to reflect the fact that most companies already have a comprehensive measurement system. The challenge is then to develop a process of transformation.

It should be emphasized that the performance measurement system is an integrated part of performance management. Satisfaction of the requirements defined in the previous chapter depends heavily on the interfaces with the other activities. We suggest the following steps in the process:

1. Business modeling
2. Develop a common understanding of priorities
3. Define external performance parameters
4. Analyze existing measurement system
5. Develop performance metrics
6. Decide how to measure
7. Collect appropriate data
8. Eliminate conflicting metrics

The two first steps are part of performance. They are included in this list to visualize the interface between planning and measurement.

Step 1 - Business modeling

Modeling is elaborated in detail in chapter 7. The performance planning model was visualized with the traditional departmental model of a company. It could easily be adjusted to different business models. A process view is necessary to develop a measurement system in accordance with the specified requirements. We will adapt the business process modeling approach as described in 7.4.1. Modeling is essential for measurement in several ways:

- describe the system and its interfaces;

- develop better understanding for cause/effect relations between activities and achievements;
- develop a cost-effective system with a minimum of conflicting interests;
- develop better understanding of operations;
- simulation of alternative solutions.

Step 2 - Develop a common understanding of performance priorities
As a consequence of the business process approach, performance priorities have to be developed for each business process instead of the company as a whole. An essential question is whether the business process model has to be integrated earlier in performance planning. If the performance priorities are regarded as universal for the whole company then it is simply a matter of transforming company priorities into business process priorities.

Step 3 - Define external performance indicators
Performance priorities have to be translated to performance indicators for each business process. This set of indicators includes stakeholders' effectiveness and strategic adaptability. They can be looked upon as external requirements and hence external performance indicators.

We will emphasize strongly that these indicators should be developed without looking at the existing measurement system. They are a consequence of performance priorities and hence requirements for competitiveness. Independence is necessary to be able to change existing values and beliefs and assure an external focus in the measurement system.

External indicators have to reveal the important characteristics of performance seen from the stakeholders point of view. An example could be product quality. Quality cost or first yield ratio are examples of internally focused indicators that should be avoided in this part of the measurement system. Real external performance indicators have to focus on quality as perceived by the customer. Examples could be acceptance ratio, mean time between failure in use and maintenance costs.

External indicators are often difficult to measure, but it is important to exclude evaluation from the creative process. Evaluation has to be carried out in a systematic way and not based on intuition. Important indicators that seem difficult to measure could be available from measurement systems by customers, suppliers or other stakeholders. This process should end up with a list of indicators covering the performance priorities.

Step 4 - Analyze the existing measurement system
As a first step, all existing metrics reported in each department or business process are listed. This registration could be combined with the priority survey, because of the close relationship between priority and measurement. Two important questions have to be answered by each department or business process:

- What is measured?
- What is reported?

The metrics from this survey should be distributed in these categories:

1. Metrics that support the performance priorities of the company or business process.
2. Metrics that accord with what is needed to fulfill these priorities.
3. Metrics that are out of alignment with these priorities.
4. Metrics that are irrelevant to the priorities.
5. Metrics that are counterproductive and likely to defeat priorities.
6. Hidden cultural-driven metrics undermining world-class performance.

Only metrics from the first two categories should be kept for further investigation. We distinguish between external and internal metrics in this process. Internal metrics are means to achieve the external metrics. This sorting could be done prior to profiling the metrics.

The remaining external metrics are compared to the list of external indicators developed previously. By comparing the two lists, we are able to detect uncovered areas and question the quality of the existing metrics. Development of new metrics should be considered if the quality is poor.

Performance measurement means an extension of the horizon in most companies. It is important to compensate this by reducing the existing measurement system. This is necessary both due to costs but equally important is the need to reduce the amount of information to manage. Traditional measurement systems have a huge number of financial indicators telling the same story and rationalization is necessary.

> In a particular company we found a 35-paged monthly report with only financial figures. This report was identical to their measurement system. They experienced large variances in the numbers from month to month and explanations where given. However, more detailed studies revealed that changes in product mix and sales volume were the main sources for the

variances. It could then be questioned how suitable this measurement system was for performance management. Rationalization was required.

Internally-focused metrics are audited against the adjusted list of external indicators to remove unnecessary metrics from the system and to reveal uncovered areas. This analysis has to be seen in connection with development of new metrics. A concurrent approach is preferred to a sequential to ensure the best match.

Step 5 - Develop new performance indicators

An analysis of the existing measurement system may reveal uncovered need for both external and internal metrics. These metrics have to be developed. In order to ensure a consistent hierarchy, external indicators have to be developed first.

We have used the term indicators deliberately to keep identification of what to measure apart from the problems with how to measure. Mixing these activities could be destructive for a creative process. Development of new performance indicators implies a decision on what should be measured.

Step 6 - Decide how to measure

Transforming all the indicators into operational metrics is a comprehensive task. Two important questions have to be answered:

- Which type of metric?
- Which measurement method should be applied?

We have previously presented a list of different types of metrics that could be useful. In most cases the type of metric is given by the nature of the indicator. The choice could be influenced by factors like:

- Purpose
- Requirement for accuracy
- Cost
- Importance
- Information available

Measurement of facts is often more costly than use of more subjective methods like evaluation, audits and surveys. This may not be true if data is available in computerized systems or may be easy to include in such systems. Most practitioners rely more on hard facts than subjective metrics. However, validity of a metric depends on the purpose and hence requirement for accuracy. If only trend is required, then cheaper methods like sampling or evaluations could be preferred to complete measurement. Cost has to be seen in connection to importance and alternative methods.

With closer relationships to stakeholders like customers and suppliers, cooperation concerning measurement could be interesting. Integration of measurement systems could help improving performance in the interfaces between stakeholder and company, which is in the interest of both.

Step 7 - Collect appropriate data

Development of a measurement system is an ongoing try-and-fail process. The final verification of a metric and the measurement method has to be done after testing the metric in use.

Step 8 - Eliminate conflicting metrics

Eliminating conflicting metrics should be an ongoing process for the company. It is easy to add new metrics to the system, but difficult to remove obsolete metrics. Research (Neely *et al.*, 1994) has shown that a formal process makes it easier for the companies to eliminate conflicts within their measurement system. Elimination of obsolete metrics is a part of a necessary maintenance of the system. Maintenance is built into the process by repeating this sequence of activities.

Elimination of conflicting metrics is integrated in the auditing processes in our approach, but we state it as an independent activity to underline the importance. It is when measurements are first available and evaluation is carried out that all consequences of metrics become visible.

9. PERFORMANCE EVALUATION

9.1 EVALUATION OF PERFORMANCE

We have distinguished clearly between measurement and evaluation in performance reviews. Reports generated from the measurement system is an important input to the evaluation process, but not the only source for information. Performance evaluation is an active process where the main purpose is to identify areas for improvement. Even if the measurement system is improved and diagnostic and competence measures are adapted, the system will still fail to go beyond business achievements and look for performance drivers. Measurement is not to confirm the level of performance, but to give maximum support to the improvement process.

The main reasons for performance evaluation could be summarized as:

- gives a realistic status of real performance and development;
- cause/effect analysis for structural and environmental changes;
- performance planning control;
- identifying performance gaps;
- identification of improvement potentials;
- rewards;
- marketing purposes.

Performance evaluation is then mainly to reveal areas for improvement. This is done partly by the measurement system, but has to be supplemented by tools that look behind the measures. As the process view becomes common, tools as self-audit, benchmarking and process analyses could be applied to give a broader picture of performance.

In a stable environment evaluation could theoretically be linked directly to the development of each performance indicator. However, as most companies are facing rapid change in requirements the situation becomes too complicated to rely on single indicators. A more balanced approach is required to evaluate real performance. Evaluation parameters could be divided in two categories; environmental changes and structural changes:

Environmental changes include:

- changes in currencies;
- changes in interest rates;
- new competitors;
- new standards;
- changes in customer preferences.

Structural changes include:

- new products;
- new technologies;
- new management systems;
- new policies.

Changes in these aspects will inevitable influence performance indicators. Feasibility of performance indicators depends on the ability to support improvement. Measurement is easy in most cases whereas evaluation could be more difficult to perform properly. Realistic evaluation is critical to justify performance measurement. Performance is a multivariable problem and evaluation has to take this into account. If the evaluation process is not trustworthy people will lose confidence in the measurement system in general.

> In a company with a performance measurement system based on standard cost we found variances up to 30%. A closer look at these variances revealed that they were caused by changes in product mix. This measurement system was used to measure productivity and people were frustrated, because variances were caused by changes in external conditions.

9.2 PERFORMANCE EVALUATION THROUGH SELF-AUDIT

In an organization dedicated to continuous improvement a need to monitor the development emerges. How can the management possibly know that the continuous change process represents a really continuous improvement process? Continuous improvement tasks within each process could lead to suboptimization and represent no real improvement on the bottom line. Traditional measurement systems are probably unable to fulfil the need for monitoring and visualizing cause/effect relations.

Future management will pay more attention to the process view. The processes have to be understood to make the right decisions both long-term and short-term. Performance of a company is the synthesis of the performance of all its processes. However, the management needs a tool for reviewing all the processes.

9.2.1 Introduction to self-audit

Self-audit is a procedure to review the performance of the business processes. We will emphasize that self-audit is a process description rather than a fixed tool available for rapid application. The notion 'self' is used to distinguish this tool from the externally conducted audits known from different consultant companies, supplier audits, quality awards and so forth. It should not be mixed up with self-control, where each employee is controlling their own work. The following collection of statements suggests some important interfaces and constraints concerning self-audit.

- Self-audit focuses on business process performance instead of departmental or organizational performance.
- Self-audit is primarily for companies devoted to continuous improvement. Better tools are available to measure effects from isolated improvement projects.
- Self-audit is a tool to manage the whole portfolio of business processes with special attention to core processes. Process analyses are better to go into detail in each single process to identify improvement potentials.
- Self-audit is complimentary to the existing measurement systems. It provides better understanding for the processes that promote better and more holistic solutions.
- Self-audit provides trends and not exact status of performance. Establishment of a reference for the indicators in self-audit could be achieved through benchmarking or customer surveys.
- Self-audit separates measurement and evaluation. Resetting goals is not a task within the self-audit.
- Self-audit is a tool to visualize development based on the assumption that 'What is watched gets done.'
- Self-audit does not include problem solving, but results from measurement and evaluation in self-audit will give valuable input.

- Self-audit is a dynamic tool, but a certain level of stability is required to be able to record development.
- Self-audit should focus on a balanced set of indicators that satisfy the needs of the different stakeholders of the company.
- Self-audit has a diagnostic purpose and identification and focus on the enablers is crucial to actively influence business achievements.

9.2.2 Development of a self-audit procedure

Self-audit is not available for rapid implementation. Each company has to tailor a self-audit scheme for the company. However, there exist some similarities for all industries making an initial procedure possible. TOPP has developed a procedure based on experience gained from different types of competitive analyses carried out in Norwegian and European mechanical industries (Moseng and Bredrup, 1993). TOPP has identified some universal business processes for mechanical industry and developed a methodology that makes it possible for a company to tailor their own self-audit. The process is divided into two phases:

- development of the first version;
- preventive and corrective maintenance during use.

Development of the first version

Development depends on the starting point for each company. A company devoted to the Total Quality Management philosophy has probably already defined its business processes and perhaps already achieved a continuous improvement process within these business processes. A typical *ad hoc* company with emphasis on short-term goals is probably not ready for self-audit. Most companies are situated somewhere between these extremes. In this section we limit our discussion to self-audit only and leave out aspects concerning process thinking itself.

- **Feasibility study**. A feasibility study is required to decide whether the company is ready for self-audit. First the company has to define the purpose with self-audit. Second, the company has to analyze the maturity regarding process thinking.
- **Collect and systemize information**. All kinds of analyses as SWOT-analysis (Strengths-Weaknesses-Opportunities-Threats analysis), market

surveys, external audits, productivity analysis and key number analysis can be useful input and help in identifying core business processes.

- **Develop the first version**. Core business processes have to be identified. The universal processes described by TOPP could be a starting point for most companies. Procedures then have to be developed for the pilot process. A pilot process provides experience to improve the development process. Internal marketing is important to achieve acceptance within the organization. Top management defines the key questions they want to be answered by the audit. Each process is ranked according to importance to decide cycle times. Indicators are then defined based on the key questions and the cycle time.
- **Application**. First step is preparation. The purpose has to be understood. Too often the need for information is underestimated. Resistance against change is considerable in most organizations and especially large ones concerning measurement and evaluation systems. Secondly, the organization of self-audit has to be defined. Collection of information could be delegated to others, but the top management has to be the driving force in the work. Self-audit is a review done by the top management and not a set of reports. Top management has to commit for self-audit concerning both own time and resources to the measurement required. Training of the people involved in carrying out the self-audit is required. Finally, a self-audit program with a detailed schedule has to be developed.

Carrying out the self-audit

This phase consists of three activities as shown in Figure 9.1.

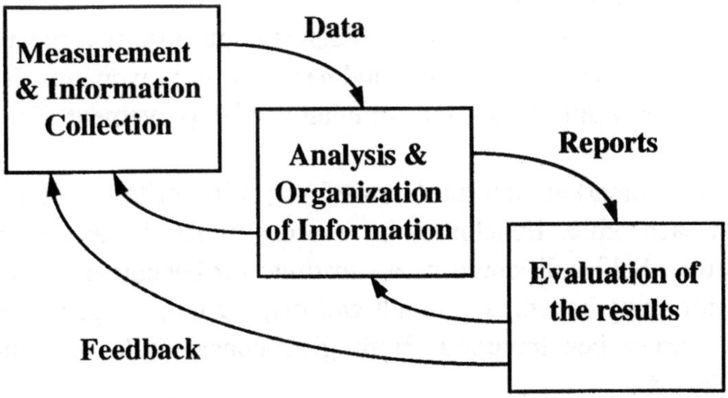

Figure 9.1 Self-audit process.

- **Measure and collect information**. Internal surveys, checklists, questions, metrics, etc., are examples of methods.
- **Analysis of the information**. Information has to be systematized. Internal surveys have to be registered in some kind of system and reports generated. Top management should be heavily involved in the specification and review of the methodology and the reports.
- **Evaluation of the results**. Evaluation is a top management activity and cannot be delegated without undermining the self-audit. It is not sufficient to passively manage reports and take decisions based on filtered information. Evaluation of a large set of indicators is a new challenge for most managers. It is neither possible nor wanted to determine an appropriate level for each single indicator. The evaluation has many similarities to medicine and methods from diagnostics have to be applied. Problem solving is not a part of self-audit, because self-audit is only one among many sources of input.

This is a simplified model. Additional information is often available especially in the evaluation phase. Interfaces to other analyses and surveys are neglected here, but have to be examined when applying self-audit. Self-audit is a supplement to a traditional measurement system and not a substitute. Coordination with other analyses is important to be cost-effective and achieve the best possible result. Elaboration of these relations is impossible within the scope of this book, but two important interfaces are visualized in Figure 9.2.

- **Benchmarking**. Self-audit provides no status due to lack of reference. Benchmarking will establish a reference for performance and is applied to improve a process and will influence self-audit due to a change of processes and new knowledge.
- **Problem solving**. Self-audit has a diagnostic purpose that means that it is supposed to pinpoint potentials and needs for improvements. It is not a question of recording status, but stimulation of improvement.

Self-audit, benchmarking and problem solving fulfill different purposes but depend on each other. Benchmarking is a costly activity and is therefore used carefully. Self-audit could be a substitute for benchmarking for those activities not being benchmarked and will help sustain the gain from those activities already benchmarked. Both give considerable information to problem solving.

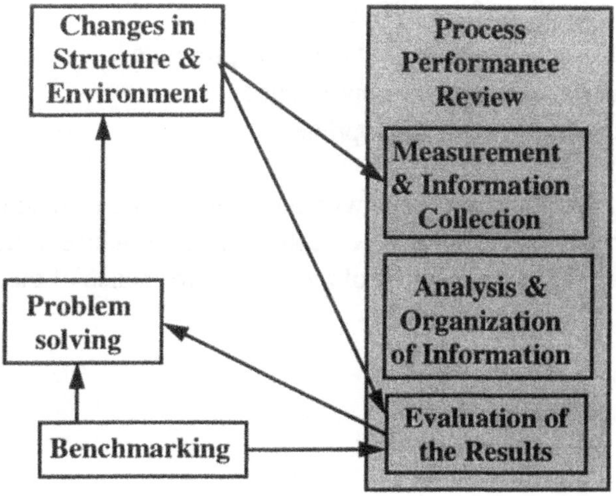

Figure 9.2 Integration of self-audit.

9.2.3 Maintenance of the self-audit

Maintenance is necessary to sustain and even improve the capability of self-audit. Capability has to be improved due to experience and changing environment. With capability we understand ability to monitor performance of the business processes. Maintenance includes:

- **Corrective maintenance** - changing and improving parts that are obviously obsolescent. This is a defensive strategy.
- **Preventive maintenance** - changing and improving parts as a response to a continuous review process, opportunities in the market, new ideas, changing environment, new information from other analyses, etc. This is an active strategy, but there exists a difficult trade-off between the request for change and the need for stability.

There are some important aspects that have to be reviewed:

- Is the portfolio of business processes appropriate?
- Are the appropriate key questions identified?
- Are the cycle times appropriate?
- Are the key questions answered by the chosen metrics?
- Are the measurement methods appropriate?

- Is the organization of the self-audit appropriate?
- Are the self-audit program (schedules) appropriate?
- Is the analysis methodology appropriate?
- Is the evaluation process appropriate?

Never forget that the main focus with the maintenance is to achieve a close correlation between the internal perception of performance in the processes and the stakeholders' perception of the total performance of the company.

PART FOUR

Performance Improvement

10. PERFORMANCE IMPROVEMENT STRATEGIES

This book has adopted the process approach due to the fact that activities are the cost drivers not the departments. An improvement of business processes has three major objectives (Harrington, 1991):

- Making processes effective - procuring the desired results
- Making processes efficient - minimizing the resources used
- Making processes adaptable - being able to adapt to changing customer business needs

These three objectives are identical to the three dimensions of performance: effectiveness, efficiency and adaptability. Improvement has to be achieved in all dimensions to ensure long-term survival. We will introduce two different but complementary approaches for performance improvement:

- Business process re-engineering (BPR)
- Continuous business process improvement (CBPI)

The differences between the two approaches are shown in Table 10.1. These methods could be combined and a typical evolution path could be illustrated by Figure 10.1.

Table 10.1 Differences between re-engineering and a continuous process.

Aspects	*Business Process Re-engineering*	*Continuous Process Improvement*
Ambition	Radical improvement	Includes small steps
Strategic link	Direct link	Not directly
Time horizon	Limited duration	Continuously
Scope	Wide	Limited
Impact	Impacts all aspects	Limited
Organization	Top-down	Decentralized
Risk	High	Low

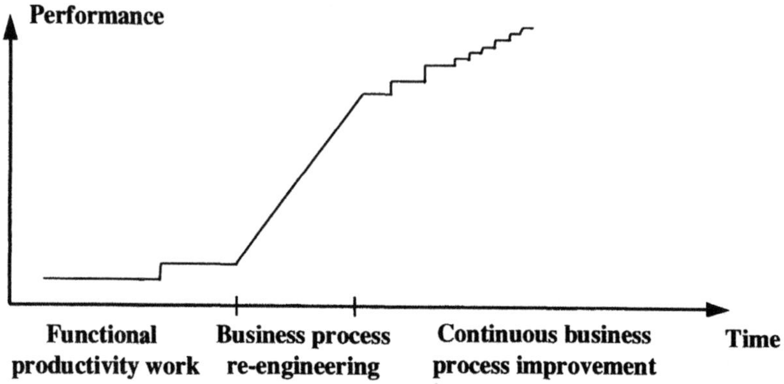

Figure 10.1 Development path with the different change methods.

10.1 BUSINESS PROCESS RE-ENGINEERING

Business re-engineering has become the new buzzword to meet intensified competition and constant need for change of increasingly complex business systems. Re-engineering could be defined as (Hammer and Champy, 1993)

> *Re-engineering is fundamental rethinking and radical redesign of business processes to achieve dramatic improvements in critical, contemporary measures of performance, such as cost, quality, service and speed.*

This definition states that re-engineering is a search for dramatic change and productivity breakthrough. Re-engineering is not the kind of tool a company applies if performance needs a 10% boost. Business process re-engineering is a strategic tool useful for achieving radical breakthroughs on business performance. Typical examples could be reduction of delivery times by 75%, reduction of cycle times by 75%, reduction of product development time by 50%, reducing the number of components by 50% and so forth. Change of this magnitude is not possible by simply improving the individual performance. That is why strong adjectives as fundamental, radical and dramatic are used to describe the magnitude of change in the processes.

It may look as if everybody is performing re-engineering in some way or another. However, true re-engineering is like starting with a clean sheet of paper. Our purpose is to present the extremities of the different approaches. In practical terms it makes no sense to operate with strict rules and most companies will operate with strategies between the extremes. A company could choose among the best tools and combine them to solve specific problems. True BPR requires vision, willpower and a comprehensive approach to change that includes these elements (Carr, 1993):

- leadership and guidance from top management;
- external focus through customer research, competitive and economic analysis and benchmarking ;
- top-level strategy to guide change and leaders who can implement change;
- methods for redesigning processes to meet performance targets;
- use of advanced information technology to enable breakthrough performance;
- effective change management and ability to develop organizational culture;
- continuous improvement methods to sustain and increase gains achieved during BPR.

This list emphasises the need for involvement from top-level strategy and top management. BPR starts with defining the outcome of the process that will best match the customer needs. This is a totally different approach compared to those who have to adjust changes to an existing process. The process is then designed to satisfy the wanted outcome. In the extreme case BPR allows the designer to start from scratch. BPR implies a dramatic change of operation and some keywords could illustrate the impact:

- **Customer orientation**. Implementations should be justified from a customer perspective.
- **Process orientation**. Re-engineering goes beyond predefined organizational boundaries and looks at the entire process to consider all tasks involved in transforming the input into output.
- **Focus on core businesses**. Re-engineering is reserved for the main processes that could translate into value and give a competitive advantage.
- **High ambition**. Applying re-engineering implicates an aim for breakthroughs, not evolution (60% or higher)

- **Rule-breaking**. No tradition or established truth is allowed to be kept out from the re-engineering process. Paradigms like specialization, sequentiality, and timing have to be abandoned.
- **Devotion for simplification**. Complexity hides waste and simplification is necessary to increase speed and reduce costs.
- **Creative use of information technology**. Creative use of information technology after simplification could enable the organization to do work in radically different ways. Re-engineering is not the same as automation. The process is the core and information technology is just a tool to reinforce improvements.
- **Rapid payback**. Re-engineering involves high risks due to the comprehensiveness and rapid payback is necessary to handle this.

It could be questioned whether re-engineering represents anything new. Some argue that the techniques behind re-engineering are known from JIT and time-based competition methods. Process flow analysis is a common denominator for several of these approaches. We will emphasise that techniques and methods in re-engineering are indeed known within current management philosophies. Re-engineering is not a streamlined concept with its own tools, but a creative approach where the result is the only matter of concern. Tools and methods are adapted whenever they suit the situation. Re-engineering is then simply a popular description for a strategy with a set of characteristics rather than a philosophy. However, people often tend to extend new buzzwords into independent philosophies.

It is impossible to describe a universal re-engineered process. However, there exist some typical sources for actions when re-engineering a process (Hammer and Champy, 1993 and Carr, 1993).

- *Several jobs are combined into one*. Benefits from combining jobs could be fewer errors and misunderstanding, less handling, increased speed, and less administration. The abolition of the specialization paradigm is supported by research (Istvan, 1992).
- *Decentralization of decisions*. Decisions should be made were the need for them arise. Decision-making becomes a natural part of the work.
- *The steps in the process are performed in a natural order*. Sequentialization forced by the organizational structure is abandoned. In re-engineered processes, work is sequenced in terms of what needs to follow what. Concurrent methods are applied when appropriate.

- *Designing flexibility into all parts of new processes.* Multiple versions of processes could be a solution. Meeting increasing demand for customization with standardization means increasing costs of omitting the standardized system and of course delays.
- *Work is performed where it makes the most sense.* Re-engineering boosts the traditional organizational structure and focus on doing the activities as close to was they are required as possible. Support activities like detailed planning, quality assurance, purchasing, etc are adjusted to make decentralization of them possible.
- *Checks and controls are reduced.* This is nonvalue-adding work and should be minimized. Re-engineered processes should tolerate modest and limited abuse, by delaying the point at which abuse is detected or by examining aggregate patterns rather than individual instances. Trust is the main rule until the opposite is proved.
- *Reconciliation is minimized.* This is done by cutting back the number of external contact points a process has in order to reduce the possibility that inconsistent data are received.
- *A case manager provides a single point of contact.* They act like a buffer between the process under re-engineering and the customer. Their role is to answer the customer's questions and solve customer problems as a shield in front of a complex process.
- *Hybrid centralized/decentralized operations are prevalent.* Advantages of centralization and decentralization are combined in the same process. Information technology and empowered operators could make hybrid organizations possible.

Most people will experience a dramatic change of work in a re-engineered business process. It will affect operators as well as managers from supervisors to top management. The characteristics of a re-engineered process described above suggests a radical change of administration and support activities in companies. This is not surprising taken into account that overhead is on average 32% of total product cost (Howell *et al.*, 1987) whereas working time is less than 5% of the throughput time. Old organizational structures will be changed and be more focused on serving the business processes of the company. This means that responsibility and authority will change. Hammer and Champy (1993) emphasize the following changes within the organization:

- *Work units change from functional departments to process teams.* Process teams actually replace the old departmental structure. It is a team that naturally falls together to complete a whole piece of work - a process. This is what Drucker (1992) calls a 'tennis double team' where players have primary rather than fixed positions. The team 'performs' and the members only 'contribute'. A 'baseball team' is the opposite were the players play on a team in fixed positions, but do not play as a team. Task flexibility is a central aspect of the lean production concept (Womack *et al.*, 1990).
- *Jobs change from simple tasks to multidimensional work.* Decentralization of planning, control and several support activities means job enrichment for people in the process teams, but also requires empowerment to succeed. Elimination of nonvalue-adding activities is a primary task and could give cost reductions or release resources to reinforce the value-adding activities.
- *Job preparation changes from training to education.* This is necessary to empower the team members because there will exist no detailed job description for them anymore. They will have to make judgments and take decisions every day. Transferring responsibility without transferring knowledge is impossible and unethical.
- *Focus of performance measures and compensation shifts from activity to results.* Time has to be given value in justification and evaluation (Borthick and Roth, 1993)
- *Advancement criteria change from performance to ability.* Advancement is change and should be a function of ability, not performance.
- *Values change from protective to productive.* People have to believe that they are working for their customer not for their bosses. However, this has to be visualized otherwise it becomes only a phrase. Bonuses could be dependent on customer satisfaction.
- *Organizational structure change from hierarchical to flat.* This is the natural consequence of the decentralization of prior management tasks. Communication flow is horizontal rather than vertical to increase speed and reduce nonvalue-activities.
- *Executives change from scorekeepers to leaders.* In traditional companies operations are managed through financial measures. However, executives have no contact with operations, but nevertheless make decisions influencing these operations. Re-engineering involves moving executives closer to the real work like for instance a football

coach. We emphasize this point heavily because the top managers' values and examples have a very powerful influence of the development of employees' values and beliefs.

We will present benchmarking and process flow analysis, which are the most important techniques in business process re-engineering.

10.2 CONTINUOUS BUSINESS PROCESS IMPROVEMENT

Continuous business process improvement is a state rather than a concept and there exist several alternative approaches and tools to reach this stage. Continuous improvement is a rather vague term impossible to define exactly. Most companies will claim they are continuously trying to improve and perhaps do so. However, is annual productivity improvement sufficient to be labeled continuous improvement or is this term connected to the use of specific tools or methodologies? We like the latter definition better than the first, but hesitate because of the danger of believing that tools are the keyword. This is mainly an attitude problem and not a matter of tools.

Research has shown that companies following the traditional productivity paradigm have achieved lower annual productivity improvement compared to companies adapting the new productivity paradigm with modern tools and methods (Miller and Roth, 1988). This clearly shows that annual improvements based on the learning curve is not enough to stay competitive. Competing with your own performance is not enough, because others will try to exploit every opportunity to beat you. Systematic search for improvement potentials is necessary. Relying on radical breakthroughs is wasted in most cases and exploiting every small or large source for improvement is the most realistic strategy.

A keyword for success is introduction of process thinking. Introducing and actually implementing the process view means totally rethinking the way the business is organized. Success depends on fulfillment of some very tough conditions:

- *Have a long-term commitment* - Developing a continuous improvement process is perhaps one of the most difficult challenges an organization could face. Change is against the nature of people, which means that resistance and obstacles are unavoidable. Without having a long-term commitment the negative forces could easily win. Most people within an organization will be of the wait-and-see-kind. It is of outermost importance to give the early adapters maximum support and to take all resistance and obstacles seriously. This war is not won with good theoretical arguments, but with powerful leadership by example and successful pilot cases.

- *Obtain management support* - Winning management support is critical because of the importance of leadership by example. This process is very fragile and middle managers could easily destroy the process. Good preparation is a keyword fully understood by the Japanese. They could prepare the process for several years to achieve absolute agreement by all managers. Agreement is perhaps one of the most important reasons for their rate of success in implementing new business approaches.
- *Focus on the process* - Traditional vertical business systems are not valid for promoting improvement of the required magnitude anymore. Horizontal operational systems are the new way of organizing established businesses for continuous improvement. These processes are the main focus for re-engineering the infrastructure. The infrastructure is developed for supporting vertical business systems and re-engineering is unavoidable. To succeed, process thinking has to be totally distributed to all members of the organization and all required activities have to be converted.
- *Assign process owners* - Choosing process owners is a critical activity, because their endurance and influencing power is essential for success. Resistance and obstacles are systematically revealed and removed.
- *Develop measurement and feedback system* - A continuous improvement process will die if measurement and feedback fail to support the process. Change is hard enough to advocate if the measurement system is unable to register improvement. Feedback is necessary to sustain the motivation and the required feeling of being on the right track.
- *Use a disciplined methodology* - This evolvement towards a continuous improvement stage has to follow a well-defined and planned road to succeed. Success depends on support and belief of all participants within the process and achieving this requires systematic work.

Description of the detailed process of achieving a continuous improvement is very comprehensive and beyond the scope of this book. Our purpose was to give a brief introduction to this matter and stimulate further studies. Harrington (1991) provides a comprehensive methodology to implement continuous improvement and he has identified five phases of continuous business process improvement (CBPI):

- *organizing for improvement* - to ensure success by building leadership, understanding, and commitment;
- *understanding the process* - to understand all the dimensions of the current business process;
- *streamlining the process* - to improve the efficiency, effectiveness, and adaptability of the business process;
- *measurement and control* - to implement a system to control the process for ongoing improvement;
- *continuous improvement* - to implement a continuous improvement process.

The linkages among the five phases of continuous business process improvement are seen in Figure 10.2. We will discuss most strategies, tools and methods necessary to design and implement a total performance management system. Our view is holistic, which means that we look upon continuous improvement as integrated in the total performance management system. We are well aware that actually implementing continuous improvement processes may require a specific methodology as described by Harrington (1991).

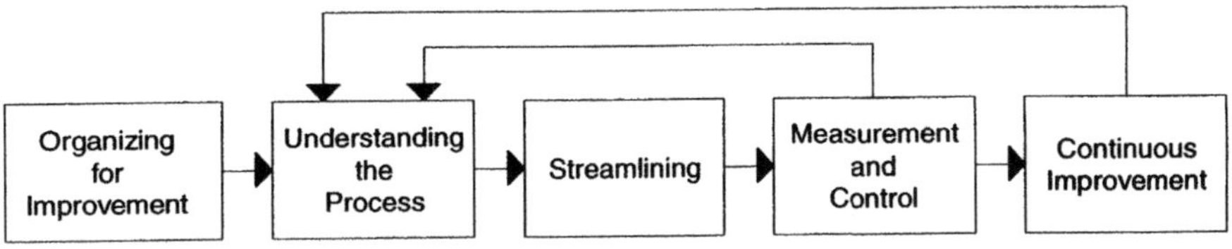

Figure 10.2 The five phases of CBPI.

11. BENCHMARKING

The core features of benchmarking are comparison and learning for improvement. Ever since the first commercially available computers were introduced, different manufacturers have made comparisons of response and processing times. When Japan started rebuilding after World War II, they soon found out that there was a lot to be learned from the more advanced American companies. Thus, during the 1950s, numerous groups of Japanese managers were to be seen touring US companies, picking up ideas. These tours were sometimes referred to as industrial tourism. These are both examples of benchmarking, to some extent.

From a management point of view, the development that has led to benchmarking as we know it today is shown in Figure 11.1. The comparison principle started when this year's results were compared to those of last year, giving some indications of how one was doing at the present time.

During the 1970s, consultants were hired to draw competitor profiles that could give even better information about the performance of one's own company. This was taken even further when Xerox introduced benchmarking as a means for setting goals, based on the achievements of excellent companies.

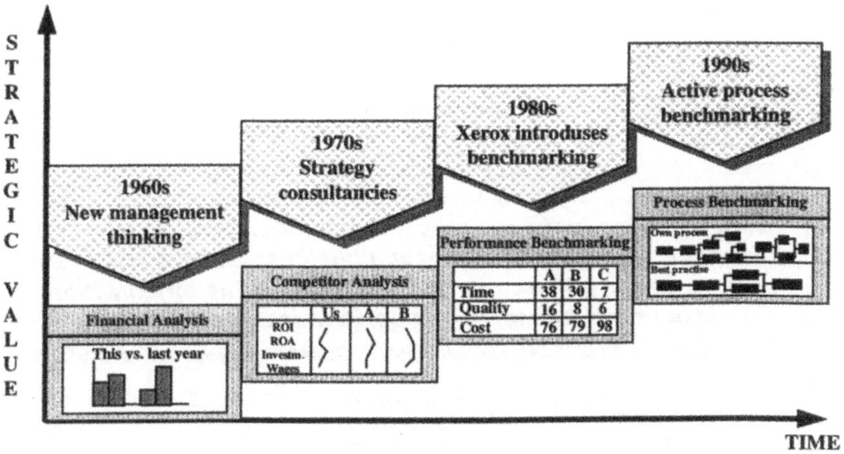

Figure 11.1 The development of benchmarking.

Since Xerox started the development of the 'real' benchmarking, emphasis has been moved from comparing performance to looking behind the performance measures to learn how to perform the process just as good as the benchmark. This is called process benchmarking, and is what we consider state-of-the-art within benchmarking at the present time.

11.1 DEFINITIONS AND DIFFERENT TYPES OF BENCHMARKING

As the field has gained interest and has been expanded, numerous different definitions of benchmarking have emerged. First of all, let us define this strange word benchmark (Oxford Dictionary = mark, standard or program used as a test or reference.):

> *A benchmark is a measured 'best-in-class' achievement, which is recognized as the standard of excellence for that business process* (APQC, 1993).

Next, benchmarking can be defined as follows:

> *Benchmarking is the practise of being humble enough to admit that someone else is better at something, and being wise enough to learn how to match and even surpass them at it* (APQC, 1993).

A more operational definition is:

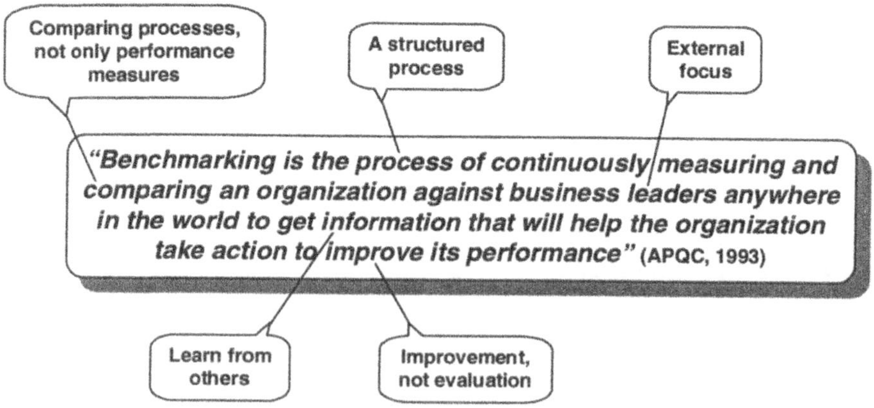

As we have pointed out earlier, the core features of benchmarking is to obtain information from others that will help the organization to improve. The focus is not to rank a number of companies through comparing performance measures, but to learn how to improve through comparing and understanding practises.

Furthermore, contrary to what many seem to believe, benchmarking is not only to compare competitors. Benchmarking against competitors is obviously one type of benchmarking, but there are others, as well. The different types of benchmarking can be classified according to two criteria:

- against whom one is benchmarking;
- what object one is benchmarking.

Depending on whom one benchmarks against and depending on what is the object being benchmarked, we separate between six different types of benchmarking, (Watson, 1992 and Andersen, 1994a):

Benchmarking against organizations

1. Internal benchmarking, comparisons between departments, plants, subsidiaries or countries within one large corporation, i.e., benchmarking against other units within one larger corporation.

This type of benchmarking is mainly applicable to corporations where processes in one part of the company are being performed better than in other parts. The less efficient processes are improved based on the better ones. Internal benchmarking can often provide a soft start into benchmarking, as access to partners is easily gained and information is readily available. Furthermore, units within the same corporation very often use the same accounting systems, the same principles of performance measures, a common, corporate vocabulary, etc. All of these elements ease the execution of internal benchmarking.

2. Competitive benchmarking, comparison of own performance against direct competitors, i.e., that produces the same products or services.

An advantage of competitive benchmarking is that competitors will often perform processes that are similar to the benchmarking company, carry similar products, compete in the same market segments, be influenced by the same external forces, etc. Due to these similarities, the validity and feasibility of the comparison is probably improved. However, when benchmarking against competitors, or rather when trying to benchmark against competitors, there can arise problems concerning confidentiality and sensitivity of data and information. The approached partner might feel treathened that he is asked to open up his operation to a competitor, while receiving nothing himself. It will therefore likely be difficult finding willing benchmarking partners for a competitive benchmarking study.

3. Generic benchmarking, this is the most common type of benchmarking, where one identifies and learns from best practise, regardless of industry. One seeks out someone performing similar processes as one self, displaying superior performance for that process. This will overcome the potential problems of competitive benchmarking, i.e., sensitive information, as well as support the sound principle of looking outside your own playground.

It is not 'roses all the way', though. Finding a company that is not a competitor and still performs a process sufficiently similar that one might learn from is not as easy as it sounds. This is a task that often requires a fair amount of creative brainstorming. There are, however, a few sunshine stories about extremely innovative choices of generic benchmarking partners:

> A manufacturer of shell cases for hand-guns experienced problems in producing cylindrical cases of satisfactory shape and with sufficiently smooth surfaces. The solution to the problems was found through a generic benchmarking study, against a partner who produced lipstick tubes, which are extremely shiny and cylindrical in shape.

Other examples are benchmarking of setup processes against Formula 1 pit crews, reduction of hospital admittance times through studying hotel receptions, improving the coordination and cooperation in a project management team by watching sail boat crews, etc.

Although not generally accepted as an independent type of benchmarking, one might define a special case of benchmarking studies, relationship benchmarking, where the benchmarking partner is a company with which the benchmarking company already has a relationship. The relationship might be that the partner is either a customer or a supplier, the two companies might be partners in a joint venture, etc.

Benchmarking based on an already established relationship can introduce some advantages to the benchmarking study:

- The benchmarking partner is already known to the benchmarking company, saving time to establish a sufficient level of general knowledge about the partner and his operations.
- The partner will generally be willing to accept the benchmarking study, as he will also benefit from any improvements his partner makes, i.e., his customer or supplier will become better and more stable.
- If the process or area being benchmarked is located somewhere at the interface between the two companies, the partner will probably as well experience direct improvements. For example, if the benchmarking company wants to benchmark purchasing, the supplier will directly benefit from the improved purchasing process of his customer.

Objects being benchmarked

1. *Strategic benchmarking*, where the objects being benchmarked are different organizations' practises on a strategic level. This includes analyzing choices of products, markets, generic strategy (according to Porter), investment levels, R&D, etc.

2. *Performance benchmarking*, comparison of pure performance levels, both economical and operational, often for the purpose of ranking a number of companies or positioning one's self against the rest of the industry.

3. *Process benchmarking*, where the objects for benchmarking are different organizations' practises, or ways of performing and organizing processes, that permit their high performance levels. Performance benchmarking will tell you how well you are doing compared to others, but only process benchmarking will tell you how to get better.

As benchmarking developed throughout the last decade and a half, there was a shift from internal and, particularly, competitive benchmarking to more generic benchmarking. Equally, there has been a shift from pure performance benchmarking to more strategic and, especially, process benchmarking. We believe these shifts to some extent reflect the potential benefits from the different types of benchmarking - a combination of generic and process benchmarking, i.e., learning to improve one's processes and performance thorough observing best practise, regardless of industry, is probably the most effective type of benchmarking. In this powerful combination, the principle of learning is very important, to such an extent that some people have started talking about benchlearning in stead of benchmarking.

11.2 WHY DO BENCHMARKING?

Benchmarking can lead to several different positive effects in the organization and the results it accomplishes. Some reasons for practising benchmarking are:

- Improving and striving for excellence through observing best practises, as well as initiating an extremely valuable quest for excellence.
- Finding ideas and sources for improvement outside one's own framework or 'box', this being either the individual company, the industry, the country, etc.
- Initiating an active learning process within the company, where one establishes a tradition for looking for and adapting good ideas.
- Creating a better understanding of one's own processes. Prior to any study of other companies, benchmarking involves thorough examination of the organization's own processes, both to understand them, and to identify which of them has the highest impact on success or has the most urgent need for improvement.
- Establishing effective strategic goals and objectives. Through learning what strategies and goals other, successful companies employ, benchmarking allows the organization to set ambitious, but still attainable goals.
- Establishing a reference point for performance measurement. If measuring the internal performance, only trend data gives meaning, the actual level of the measures does not give any information. By using external benchmarks as reference points, the level of the performance measures will give additional meaning.
- Creating a sense of urgency. Discovering the results other companies are attaining, which are often decades ahead of your own, a sense of urgency is immediately born within the organization. This might very well help

shake the company out of a state of complacency, where there is no feeling of need for continuous improvement. This feeling has proven to be fatal, see for example Dertouzos, *et. al.* (1989).

- Understanding and meeting changing customer requirements. By going through the process of identifying the organization's critical success factors in satisfying its customers, the organization will gain valuable insight into what processes must be improved. In the next phase, learning from others helps understand how to improve these processes in order to meet the customer's requirements.

By the way, there are different theories about the origin of the word benchmarking. According to Webster's dictionary (1990), it comes from geographic surveying, where it means:

> *Benchmark, a marked point of known or assumed elevation from which other elevations may be established.*

Another theory claims that the word originated from the sale of fabrics, where the shops had, and still have, marks or a ruler mounted on the counter or bench and took measures by benchmarking the fabric against the ruler.

Still others believe the word to come from fishing contests, where one compared the size of the fish by benchmarking. The first fish was laid on a bench or counter and its size recorded by making a mark in the counter at the end of its tale fin. By putting on the next fish and checking whether it reached the mark or not, one could establish the winner of the contest.

1.1 WHAT IS ACHIEVED THROUGH BENCHMARKING?

While we are strongly promoting the use of benchmarking, we are in no way claiming that benchmarking should be the only means for improvement. Benchmarking should be one of several tools available to the company for driving improvement. By this, we will strongly emphasize that benchmarking is not supposed to replace efforts for continuous improvement, but rather be a supplement.

Hayes and Wheelwright (1984) and Imai (1986) contrast the two approaches of improvement through strategic leaps, or innovation, and continuous improvement. Strategic leaps often happen as the result of making an all-out effort, initiated by management and carried out by R&D

people. This approach does not necessarily require low-level expertise trained in improvement techniques. Just putting all heads together as an act of shear necessity can produce the desired results, although it often happens that the necessary breakthrough is not achieved.

Working on a strategy of continuous improvement, however, requires trained employees at all levels contributing to the organization's low-level expertise. An organization possessing these qualities and being able to produce improvements continuously, is also capable of once in a while coming up with a breakthrough improvement as well. It is thus much safer to employ the principle of *kaizen* and harvest the occasional breakthrough improvement, than to depend on necessary strategic leaps at critical points.

The interesting point about benchmarking is that, as the organization continuously improves, benchmarking can effectively add the seeds for breakthrough improvements. Combining *kaizen* and benchmarking can give the benefits from incremental improvements and the killing punches of occasional strategic leaps, as shown in Figure 11.2.

When it comes to numerical examples of what can be achieved through benchmarking, we will present some by the use of an example:

> A company benchmarking their purchasing processes and supplier relationships, found the following differences between themselves and the benchmarking partner:

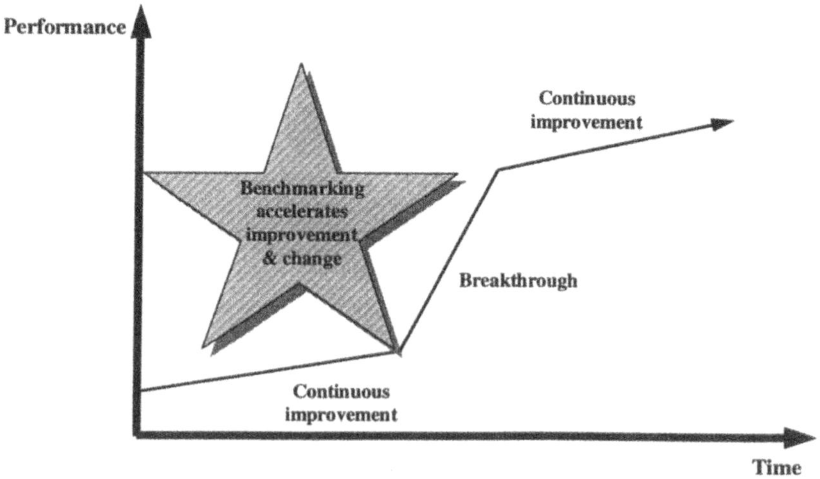

Figure 11.2 Benchmarking can effectively supplement continuous improvement.

	Company A	World-class
Lead time from suppliers	150 days	8 days
Order reception time	6 minutes	on-line
Too late deliveries	33%	1%
Number of out-of-stock/year	400 times	4 times
Suppliers/customer	34	5.3

In addition, the world-class company actively involves the suppliers in the product development process.

11.4 PRECONDITIONS FOR BENCHMARKING AND WHEN TO USE BENCHMARKING

Benchmarking is not a suitable tool in every company or every situation. In both cases, there are some conditions that should be satisfied before one decides to benchmark, (Andersen and Pettersen, 1994a).

The company-related conditions can be split into three categories:

1. Structural conditions, i.e., resources and abilities needed to carry out benchmarking should be present in the company. These are:
   - financial resources, among other costs, certainly to finance site visits;
   - time, to do all tasks necessary for a successful benchmarking study;
   - competencies, quite simply sufficient knowledge about benchmarking to be able to do it;
   - competitive abilities and potential for development;
   - documentation of central processes.

2. Cultural conditions, i.e., the values and attitudes that forms the basis for effective use of the benchmarking principles must be present in the organization. These are:
   - international aspiration;
   - will to change;
   - will to information sharing (both internally and externally);
   - management commitment;
   - participation of the employees.

3. Understanding of and consciousness about the company's business processes, as business processes are the main unit of analysis in a benchmarking study. The following dimensions should be present:
   - process documentation, e.g., flow charts;
   - an understanding of how the different business processes impact on competitiveness and critical success factors;
   - performance measures for the business processes.

If all, or at least many, of these conditions are satisfied, benchmarking is probably a suitable tool for the organization. But not in every situation, see Figure 11.3.

As the figure shows, benchmarking is the technique that demands most resources and time to carry out, while at the same time offering the highest potential for improvement. Benchmarking should therefore be used:

1. For the business processes that have the highest impact on the critical success factors.

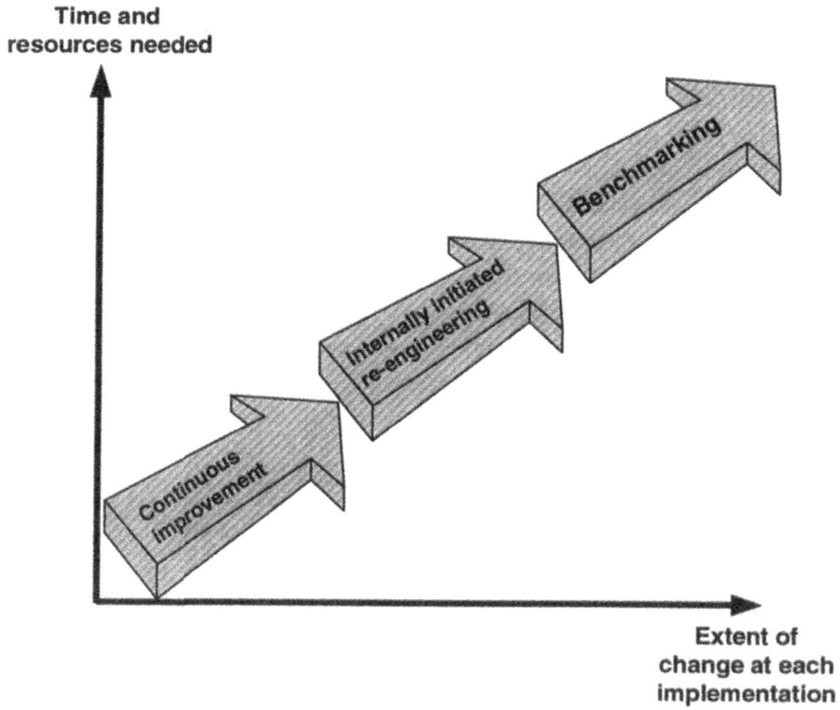

Figure 11.3 When to use benchmarking.

2. When the company has the necessary time and resources (as already described).

In addition, benchmarking should:

- be linked to the company's strategy and strategic planning;
- be linked to the general problem solving and improvement processes of the company;
- constitute an integrated element in the company's TQM-efforts.

11.5 THE BENCHMARKING PROCESS - THE BENCHMARKING WHEEL

Different sources, i.e., companies and academic institutions, have designed numerous different benchmarking process models. All of these vary highly in the number of steps, the shape, how the continuousness of benchmarking is illustrated, etc., but the basic contents of all the models are quite similar:

1. Choose and document the process to be benchmarked.
2. Identify who performs this process better, preferably, best.
3. Observe how and how well the benchmarking partner performs the process.
4. Identify the causes for the superior performance of the benchmarking partner.
5. Implement improvements in one's own process based on the information obtained from the benchmarking partner.

After studying quite a few of these different process models, we feel that a circular shaped model, with the suitable number of five steps, represents a model that is easy to understand and use. We have called this model the benchmarking wheel, see Figure 11.4, (Andersen, 1994b, Andersen and Pettersen, 1994a, 1994b).

11.5.1 The planning phase

The planning phase is both the phase that probably will take the longest time to complete, as well as the most important phase of the entire

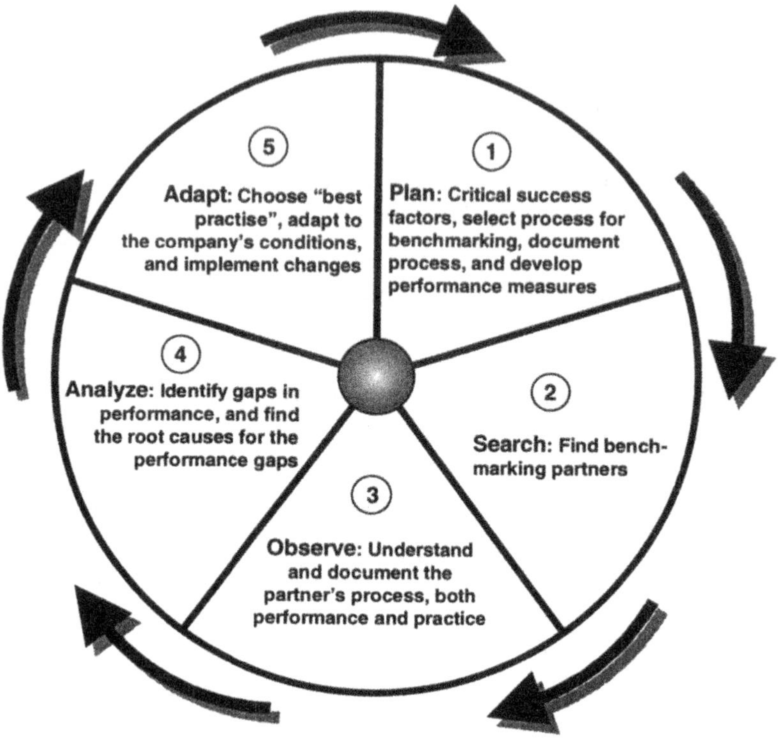

Figure 11.4 The benchmarking wheel.

benchmarking study. The foundation for an effective and successful study is laid at this point. It might be tempting to rush the planning in order to reach the 'real benchmarking', i.e., visiting partners, but the site visit is bound to be a failure if one turns up unprepared.

The tasks to be performed in the planning phase are as follows.

Choose the process to be benchmarked

The process can be chosen according to three main criteria:

- The process' impact on the company's critical success factors.
- The process' importance with regard to the main functions of the company.
- Processes that are or influence problem spots in the company.

These criteria will help to ensure that the area chosen for benchmarking really is important, so that one avoids wasting precious resources on unimportant processes, see Figure 11.5.

Form benchmarking team

The benchmarking study should be carried out by a benchmarking team, formed on an *ad hoc* basis, especially for this study. The team needs to fill the following roles:

- team leader;
- management representation or connection;
- process owner(s);
- people involved in the process (also from adjacent departments that the process runs through);
- customer to the process, internal or external (if possible);
- facilitator, i.e., an internal or external consultant that can guide the team.

One person might represent several of these roles; the requirement is that the team together can fill them. The ideal size of a benchmarking team is three to six persons. If there is a need for even more people, one solution is to call in the necessary persons when needed, without appointing them

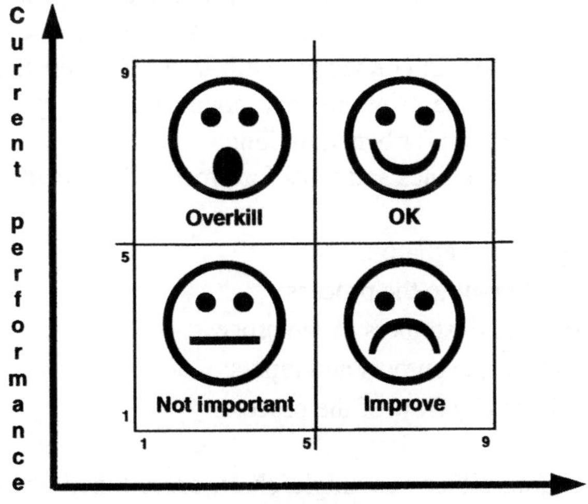

Figure 11.5 Performance matrix for identifying what processes to benchmark.

permanent members of the team. This way, they will constitute a peripheral team.

When putting together the team, there are some requirements that should be satisfied by the potential team members:

- time to participate actively in the study;
- competencies and knowledge, both about benchmarking and the company and process at hand;
- motivation and desire to work with benchmarking;
- team player with the ability to listen and communicate;
- credibility and respect in the organization.

The worst strategy for recruiting benchmarking team members is based on the thinking 'You have nothing else or better to do, go join that benchmarking team!'

Understand and document the process to be benchmarked

This is an important part of doing the homework of the benchmarking study. Fully understanding one's own process is an absolute prerequisite to being able to spot ideas for improvements in the partner's process.

As process management and other management philosophies based on horizontal processes, as TQM - Total Quality Management, ABM and ABC - Activity Based Management and Activity Based Costing, TBM - Time Based Management, BPR - Business Process Reengineering, etc., are gaining interest, more and more companies have already documented their main business processes. In connection with ISO 9000 certification, others have developed process descriptions that might serve as a good foundation for process documentation for benchmarking.

There are a few guidelines and tools for process documentation:

1. Describe the process verbally:
   - suppliers and input to the process;
   - output from and customers of the process;
   - requirements to the outputs and inputs;
   - the internal activity flow of the process.

2. Draw flow charts, either one single chart, cross-functional charts, flow charts on several levels, etc., depending on what have been used in the organization earlier. Flow charts are treated in chapter 13.

When drawing the flow charts, the whole benchmarking team should naturally be involved, but this is one of the activities where it may be wise to include other persons as well. To ensure that the flow chart is as accurate as possible, all departments and teams involved in the process should review the charts. If the flow chart is based on already existing process documentation, make sure that the chart depicts the real process, as it is performed today, not an ideal situation the way it should be.

Establish performance measures for the process
Performance measures are necessary to:

- Know where we are today with regard to performance levels.
- Compare own performance levels with the performance levels of potential benchmarking partners.
- Assess the improvements achieved through benchmarking.

The development of performance measures and what types of measures could be used is treated extensively in Parts II and III of the book. We will therefore just briefly mention the three main dimensions of process performance:

- Quality, often measured as defect rates or the process output's ability to satisfy the requirements.
- Time, generally time to execute steps in the process or time to deliver outputs from the process.
- Cost, either cost of performing steps in the process or cost of outputs from the process.

The performance measures used for benchmarking purposes can also be divided into two categories, depending on what the measures describe:

- Result measures, that describe the result of the process, often the quality, time or cost of the final output.
- Process measures, that describe how the process is performed, e.g., times for different sub-steps of the process, how many people are involved, equipment used, etc.

Considering the last type of measures, process measures, that should describe the way the process is performed, one also quickly realizes that the performance measures do not always have to be quantitative. In order to describe, for instance, what tools or equipment are used to carry out a process, it may be necessary to use qualitative measures as well.

When the performance measures have been established, the next step is to gather information on the performance levels of the process. If the data needed to calculate the measures have been collected earlier, one can easily calculate the performance for some time back. This is useful, as one is also able to see the trend in performance. If the data needed has not been collected before, one needs to take some time to collect it to establish the current performance level of the process.

When all of these activities have been completed, the crucial planning phase has reached completion.

11.5.2 The search phase

After the process to be benchmarked has been chosen, based on that process' needs, a benchmarking team has been formed, the team has understood and documented this process, as well as developed performance measures for it, the time has come to identify benchmarking partners. After completing the search phase, the benchmarking team should have identified a list of potential partners, selected for their superior performance level for the process being benchmarked, and made contact with a number of these for acceptance of participation in the study.

The search phase can generally follow these steps:

1. *Make a list of criteria a benchmarking partner should satisfy.* An apparently excellent benchmarking partner may not after all be suitable, due to several different conditions, for example:

    - geographical position;
    - size;
    - structure and organization;
    - products;
    - technology.

    By setting up such a list, one can ensure that non-suitable partners are removed from the potential partner list as early as possible, and before much effort is put into making contact with the company.

2. *Identify potential partners.* There are numerous different sources of information that might indicate that companies are good at different areas:

- Public information, e.g., national statistical agencies, OECD, export counsels, trade offices, etc.
- Trade and industry organizations, which can provide statistics and reports, conferences and seminars, fairs and exhibitions, etc.
- Publications, e.g., journals, magazines, books and indexes, company catalogs and annual reports, product catalogs, data bases, etc.
- Trade and area experts, e.g., academic institutions, research centers, consultants, etc.
- The company's own network, e.g., distributors, customers, competitors, suppliers, strategic partners, benchmarking clearinghouses, etc.
- Special attention, e.g., media attention, quotations, award winners (quality awards, etc.), etc.

3. *Choose partner(s).* All consulted sources will probably have suggested numerous companies that are good at the particular aspect that has been selected for benchmarking. Based on their performance levels and the degree to which they satisfy the criteria for benchmarking partners, the potential partners should be ranked by how preferred a partner they are. The prioritized list is the basis for selecting partner(s). The reason why we place the plural 's' in parenthesizes is that in some cases, having only one partner may be preferable. However, generally we would recommend using more than one partner, as this will provide more information and a possibility for cross-checking information between the partners. Taking into account the cost and time side of the equation, three partners seems an ideal number.

4. *Establish contact with the selected partner(s)* to request their participation in the benchmarking study. Contact can be established either by phone or with a letter. Remember to present yourself and the company you represent, describe the benchmarking study and what process you want to improve, and consider enclosing examples of questions you want the partner to answer.

People new to benchmarking are very often skeptical that no company in their right mind will accept visits from benchmarking companies. These people are very often surprised by the partners' curiosity and willingness to share information.

Why do so many companies participate in benchmarking studies? There are several reasons:

- Curiosity; a mutual curiosity for seeing what other companies do and how they do it, even though the company already is good.
- The possibility for learning; not necessarily within the area one is asked to share information, but within other areas where the benchmarking company is good. In fact, many companies learn benchmarking through being benchmarked themselves.
- Information in return, as there is an element of give and take in benchmarking. By giving away information, one expects to get some information in return, either as a report from the study, information about how the benchmarking company presently performs the process being benchmarked, or as a 'raincheck' for benchmarking the partner at a later stage.
- Professional contact; in the event of a benchmarking study, contact is established between professionals in similar jobs, but in different companies. This type of contact is interesting and educational.
- Politeness; many companies will participate in a benchmarking study out of pure politeness. If the request is put forward in a polite and decent way and it seems that the benchmarking company has done its homework, most companies will try to comply with the request, at least within reasonable limits.

11.5.3 The observe phase

The objective of the observe phase is to study the selected benchmarking partner(s) to understand their process, in order to learn enough to becoming better. To be able to utilize the information obtained from the partner, information should be obtained at three levels, see Figure 11.6:

1. Performance level, that indicates how good the partner is compared to ourselves.
2. Practise, the way the process is organized and performed that makes it necessary to achieve such performance levels.

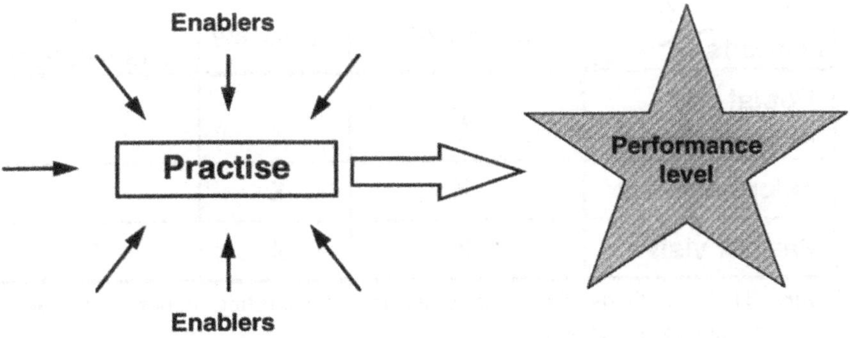

Figure 11.6 The relationship between performance level for a process, the practise of the process, and enablers.

3. Enablers, conditions of the organization or the surroundings within which the process is working, that permit the practise that allows the achieved performance levels. Enablers could be:

- training;
- organization of the process;
- equipment in use;
- computer systems;
- communication;
- support processes.

One should keep in mind that much of the information about the partner can be collected from other sources than the partner himself. External, secondary information like annual reports, financial reports, media coverage, statistics, etc., can generally provide much of the necessary background information about the partner. By taking advantage of these sources, the partner visit will be far more efficient, both because time is saved not having to go through the general presentation of the company and because the benchmarking company is better able to understand the partner's processes due to their knowledge of the partner.

There are several different ways of collecting the desired information from the benchmarking partners. To be consistent, we separate between methods and tools for this purpose. By method we mean the method for contacting the partner, while by tool we think of the specific tool used in connection with one or more of the methods. This might sound a little confusing, but hopefully the matrix in Figure 11.7 will make things clearer.

Methods \ Tools	Questionnaire	Interview	Direct observation
Postal investigation	X		
Telephone	X	X	
Partner visit	X	X	X

Figure 11.7 Methods and tools available for collecting information from the benchmarking partner.

The crosses indicate what tools can be used in connection with the different methods.

The different methods all display some advantages and disadvantages:

Postal investigation, advantages:

- Inexpensive.
- The respondent can answer the questions at his own pace, when he has the time, and can consult sources if necessary.
- Information that would not be readily available during an interview or partner visit can be collected.
- The respondent can remain anonymous.

Postal investigation, disadvantages:

- Low response rate.
- Problems in interpreting both questions and answers.
- Information can be filtered.
- Probing deeper into questions that are unclear or insufficiently answered is difficult.

Telephone contact, advantages:

- Relatively inexpensive.
- Easy to plan and execute, and can be carried out wherever and whenever.
- Easy to contact large numbers of persons.
- Can be used for detailing already gathered information.

Telephone contact, disadvantages:

- Hard to get in touch with the right person.
- Telephone contact is often very impersonal.
- Difficult to explain questions and answers over the phone, mainly due to the lack of visible contact.
- One is easily interrupted.
- Problem getting people to return calls.

Partner visit, advantages:

- A partner visit allows a degree of detail and depth of information that is impossible to obtain through the other methods.
- During a partner visit, one can oneself study best practise in action.
- A good foundation for a more long-term relationship to the partner.

Partner visit, disadvantages:

- Expensive.
- Takes much time, both to plan and execute.

Some general advice about the observe phase:

- Plan in advance what information is wanted, as it both makes the collection process more efficient and effective as well as giving a better result in the end.
- It is generally more useful to collect detailed information on one focused area than to spend much time on collecting general information about the partner. The general information should be examined during a pre-study of the partner.
- After each data collection session, the team should perform a debriefing session to summarize what has been learned, to check whether all areas have been covered, and to ensure that everything has been taken down and made notes of.

11.5.4 The analyze phase

The objectives of the analyze phase are to identify gaps in performance between own and partners' processes and the root causes for these gaps, confer Figure 11.6. The analysis includes the following steps:

Normalization

A common argument against benchmarking, is:

> *'We are so different from everyone else that any comparison against others is a waste of time.'*

However, most things can be compared, but for the comparison to be valid, there is often a need for normalizing the data to be compared. The normalization is performed in order to make up for factors that would otherwise make the comparison incorrect. In stead of uncritically comparing apples with pears, one tries to convert the pears to something equivalent to apples.

The normalization will often consist of converting data to per year, per employee, per etc., in order to validly compare numbers that might change with the size of different factors. Factors that render normalization necessary could be:

- Content of the company, i.e., what tasks and processes the company performs.
- Size of the company, e.g., number of employees, turnover, etc.
- Degree of vertical and horizontal integration, e.g., are the distribution channels integrated in the company.
- Market conditions, with regard to geographical position, size, behavior, expectations, etc.
- Cost structure, due to property prices, government support, wage levels, etc.
- Age of the company, which often impacts mode of operation, culture, attitudes, etc.
- International differences, due to different rules and regulations, trade laws, country-specific conditions, etc.

Identification of gaps in performance

Through comparing performance levels for ones own and the partner's process, we can identify gaps. It is, however, not only a momentary picture of the gap that is of interest. We are also interested in the development of this gap over time, the trend, see Figure 11.8.

The easiest way to compare performances for the sake of identifying gaps, is to simply compare performance matrices for the companies. To compare the trends, graphs like the one in Figure 11.8 can be drawn. There are a couple of more sophisticated ways of identifying gaps, but we will not present these here. We feel that this is nothing more than a simple mathematical operation that need not be explained in any more detail.

Identification of the causes for the gap

There are several different tools and techniques for this exercise, from simply comparing flow charts to spotting differences in the processes, to using cause-and-effect diagrams and root cause analysis (which are parts of the general TQM toolbox) to the more specialized relation diagram. The traditional TQM tools are described in chapter 13.

If comparing flow charts, one should look for:

- steps that are present in only one of the processes;
- loops that are present in only one of the processes;
- different grouping of tasks into steps;
- different order of steps;
- different organization of steps, e.g., parallel versus serial;
- different connections between steps.

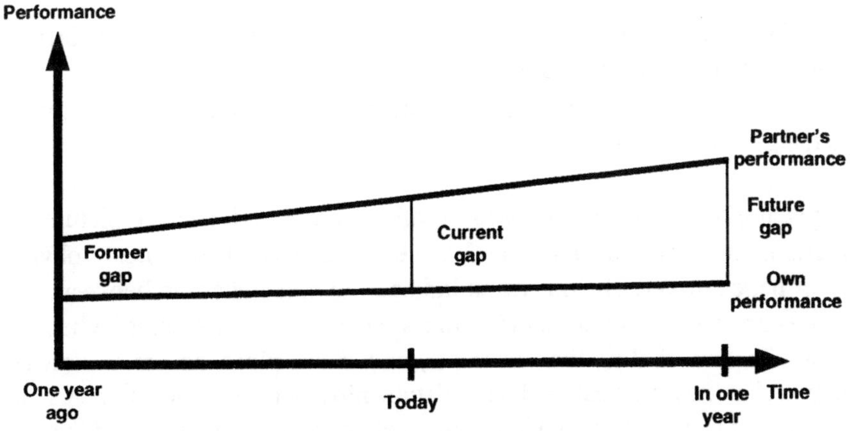

Figure 11.8 The development of the performance gap.

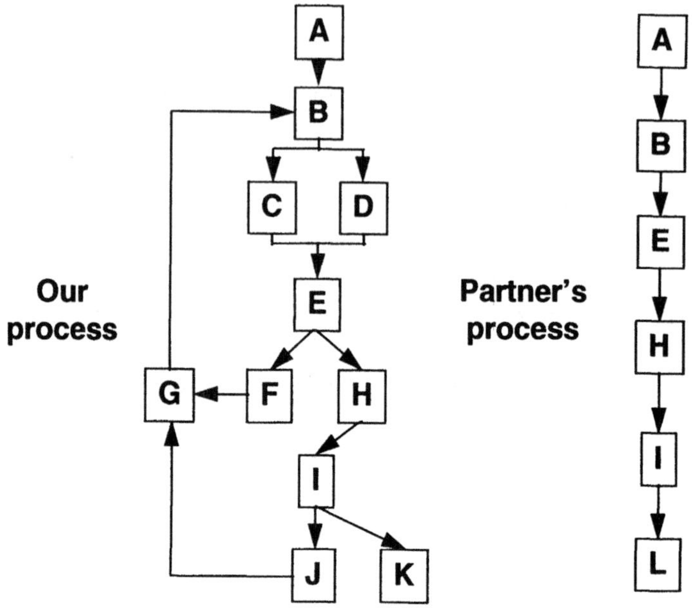

Figure 11.9 Comparing flow charts to identify differences in practise.

An example of a flow chart comparison is shown in Figure 11.9.

If we have measured several different performance measures for the partner's process, and suspect that there are some relationships between these, the relation diagram is an effective tool for resolving the directions of impact, APQC (1993). In the performance picture, the elements, or performance measures, can generally only be one out of two different functions:

- Performance driver, i.e., having an impact on or being an enabler for other performance measures.
- Result indicator, i.e., that shows or indicates a result as a consequence of changes in a performance driver.

By placing all elements supposed to influence the performance picture in a diagram, as shown in Figure 11.10, we can analyze their roles - obvious performance driver, obvious result indicator or somewhere in between.

To construct the relation diagram for each element, we assess whether it influences any of the other elements, and if so, place an arrow between them. After having analyzed all relationships between the elements, the number of arrows going in and out of an element is counted. A performance

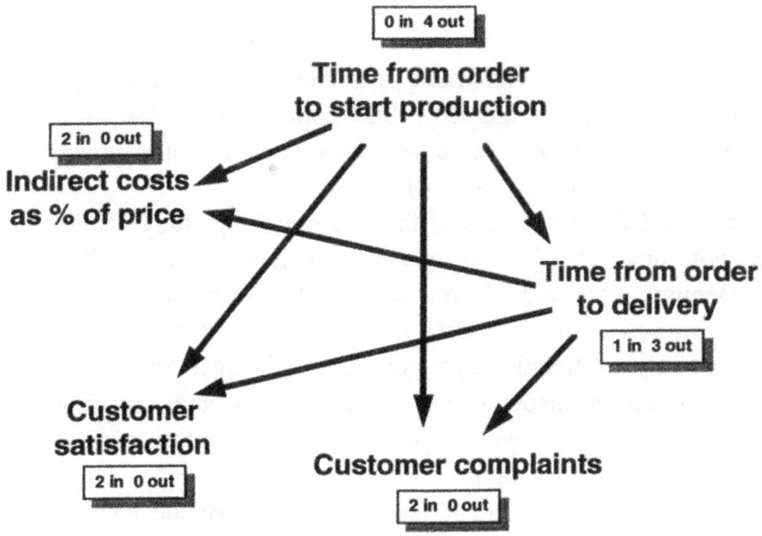

Figure 11.10 Example of a relation diagram.

measure with several arrows pointing out of it, i.e., more than pointing in, is probably a performance driver. Accordingly, a measure with many arrows pointing in, is probably a result indicator.

In the example diagram shown in Figure 11.10, the time from order to start production and the time from order to delivery are the main performance drivers, while the result indicators are customer complaints, customer satisfaction and indirect costs as a percentage of price.

When trying to identify the root causes for the performance gap, we have to start with the performance drivers. They are the factors driving the process and creating the obtained performance level.

Through the analysis, several different gap causes will generally be identified. To what degree each of these causes contributes to the resulting performance gap is often determined by the Pareto principle, i.e., 20% of the causes contribute to 80% of the gap. As we rarely will have the necessary resources or find it practically convenient to implement all of these in our own organization, we should, as a last activity in the analyze phase, rank the identified causes according to importance. This ranked list will form the basis for the last phase of the benchmarking study, where we will plan and execute the adaptation of the practises and enablers that will lead to improved performance.

11.5.5 The adapt phase

The ultimate goal of benchmarking is to create changes and improvements. Benchmarking includes elements like comparison, information sharing, analysis, etc., but unless these activities lead to improvement, the full potential of benchmarking has not been exploited.

Notice that this last phase of the benchmarking study has been named adapt, not copy! This is also very consistent with the words of the late W.E. Deming, who said 'Adapt, not adopt!' After having observed best practise, in action and working in the operations of the partner, it is tempting to try to directly copy this into one's own organization. Unless one's organization displays the same enablers and conditions as the partner's, this strategy will almost certainly fail. We must adapt best practise to our own conditions.

The first step of the adapt phase is to choose the elements in best practise that we want to proceed with. It is very likely that we do not have the necessary resources to adapt every aspect of the partners' processes. We must therefore choose the critical few that will give the highest effect. The selection can be based on:

- the process' adaptability to one's own organization;
- investment requirements;
- training requirements;
- financial resources necessary;
- time limits on the project;
- team motivational level.

The next step is to set goals for the coming improvement efforts, both based on the partners' achievements and based on what is believed to be theoretically possible in our own organization. The goals should be both ambitious, based on the performance of the best, while at the same time realistic, related to the potential of one's own organization. It is also important to integrate and coordinate the goals set in the benchmarking study with the improvement goals of the rest of the organization. One method for projecting goals based on the partners' performance levels, is gap analysis, APQC (1993).

First, we indicate the partner's and our own performance levels in a performance diagram, see Figure 11.11.

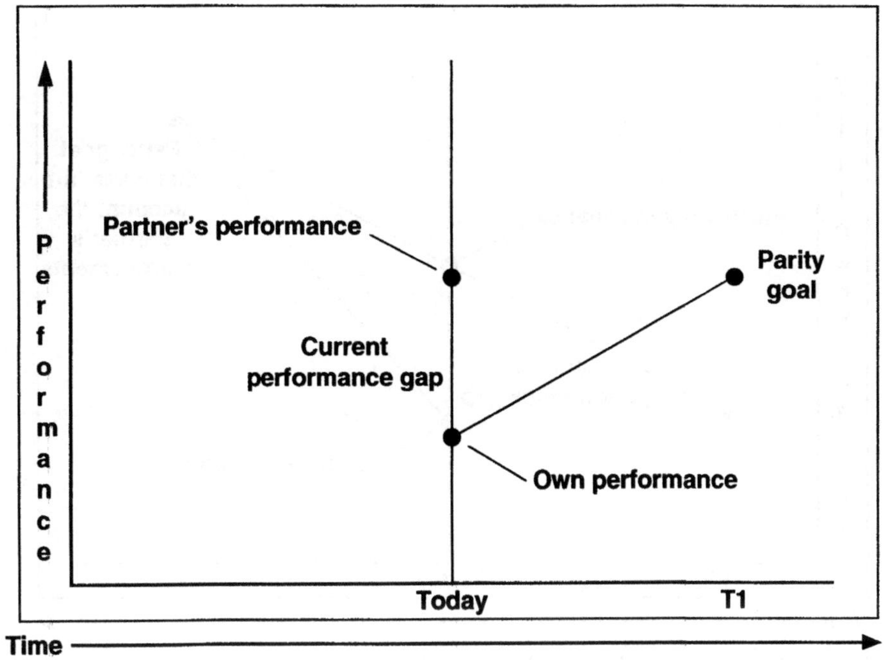

Figure 11.11 The first step of the gap analysis.

Considering only this information, it would be logical to appoint the partner's performance level as the goal for our own improvement. We would then close the gap and have achieved a performance level at parity with best practise.

However, if we are able to look back in time, we will probably discover that since time T1, the partner has improved a certain amount, see Figure 11.12.

It is natural to assume that the partner will continue to improve, and probably somewhere near the same rate of improvement. This means that at time T1, the partner will have reached a much higher performance level. Based on the argument that someone else has achieved it, this level will also represent a realistic goal for our own improvement. Our goal should therefore be to achieve parity at time T1, which will require us to improve faster than the partner.

When the specific improvement ideas have been chosen and goals have been set, the actual benchmarking part of the study is completed. From now on, the process is similar to any general implementation process of improvements, including activities like project planning, plan implementation, monitoring progress and reporting. These activities have

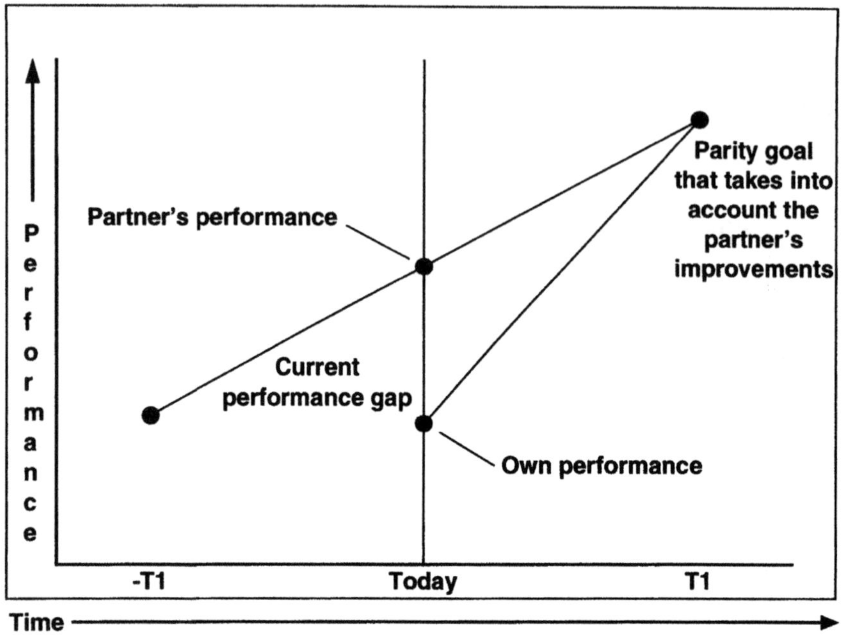

Figure 11.12 The second step of the gap analysis.

been covered in other sections of the book, therefore we will just briefly mention the steps required:

1. *Develop a project plan for the implementation of best practise.* The plan must cover:

- Activities necessary to implement best practise.
- Activity sequence, i.e., the order of the activities.
- Organization, i.e., who performs what activities.
- Schedule, i.e., when are the activities to be performed.
- Costs, i.e., a cost estimate for the project.

With regard to organization of the implementation activities, there are basically three different ways to choose from:

- Implementation by the benchmarking team, where the same team that has carried out the information collection and analysis also implements the changes.

- Implementation by an improvement or implementation team, where another team is formed that is responsible for implementing the changes identified by the benchmarking team.
- Implementation in the line organization, where the regular organization implements the changes, pulling on the resources available in the organization.

To maximize the likelihood of success of the plan implementation, it is wise to try to get as much support for the plan as possible. A useful technique for analyzing the organization's attitudes towards change, is force field analysis. Force field analysis is based on the assumption that any situation is a result of forces for and against that balance each other. If either of the forces change in strength, change in the situation will occur. Prior to starting the implementation of the project plan, a force field analysis should be undertaken, to identify the forces working for and against the plan. These forces are depicted in a diagram, as shown in Figure 11.13. Note that the length of the arrows indicate the strength of the forces.

When one has acquired an overview of the force situation, this information must be utilized to ease the task of implementation. This is done through applying the following to simple principles:

- Strengthen the forces working for the change.
- Reduce the forces working against the change.

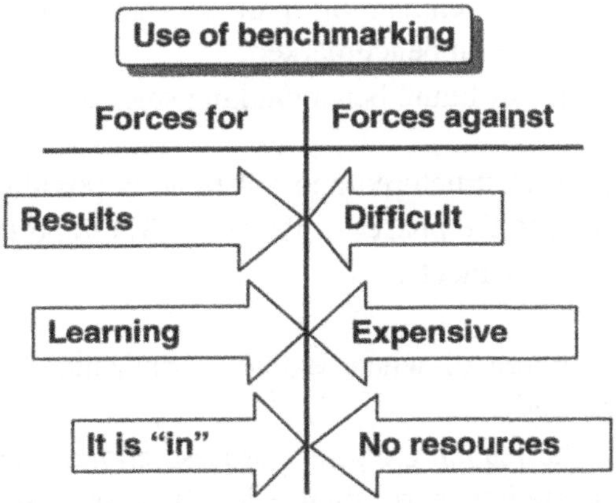

Figure 11.13 Force field analysis of the use of benchmarking.

2. *Implement the project plan.*

When the best possible foundation for the implementation plan has been laid, through a sound plan, a sensible choice of organization of the project and support for the plan, the time has come to implement the improvements. We have no quick fixes to recommend here - thorough preparations and common sense are the best ingredients for success.

3. *Monitor the progress of the implementation.*

To ensure that the project stays on track, meets the set goals, and stays within the projected costs and time limits, the actual progress must be compared with planned progress. Deviations from the project plan that might occur must be corrected.

4. *Final report from the benchmarking study.*

The last activity in the last phase of the benchmarking study, is to document the study in a final report. This report will serve several different purposes:

- Description of the project to all stakeholders.
- Documentation of the analysis, that might serve as a basis for future improvements.
- A summation of the project to the benchmarking partners.
- A file detailing the information and data gathered and analyzed.
- A file describing benchmarking partners and contact persons with processes that have been benchmarked.
- Recommendations for future benchmarking studies.

To make the report as structured and coherent as possible, it can then be structured according to the phases of the study. The report should generally contain the following elements:

- A project description of when, extent, organization, etc., i.e., a brief summary of the study.
- A summary of the planning phase - process chosen, team composition, process documentation, performance levels before improvements were made.

- A summary of the search phase - benchmarking partners evaluated and selected.
- A summary of the observe phase - partners visited, information gathered
- A summary of the analyze phase - results, identified improvement opportunities.
- A summary of the adapt phase - specific improvement projects, project plans, improvements achieved.
- Experiences from the study and recommendations for future studies.
- Identified improvement opportunities that were not implemented or further developed.

11.6 SOME FINAL IMPORTANT POINTS ABOUT BENCHMARKING

- Benchmarking includes both quantitative and qualitative data and analysis, i.e., both performance levels and how these levels are achieved.
- Comparison of key figures, often collected by external consultants, is not complete benchmarking.
- Just being presented performance levels, without participating in the collection and analysis, often leads to the three D's: Disbelief, denial and despair.
- Effective benchmarking requires ownership of data and information, both through development of the study and collection of the information.
- Benchmarking needs to be a continuous process - there will always be new areas to benchmark and existing benchmarked areas will be improved by others.

Please, do also remember, and avoid, the common barriers against effective benchmarking:

- The fear of being perceived as copy cats, which forces many to reinvent the wheel, every time.
- The fear of giving away competitive advantage through sharing information.
- Arrogance - 'We are best and don't need to benchmark!'
- More arrogance - 'Benchmarking is so easy, you just go ahead and do it. We don't need to learn any structured or systematic method!'

- Benchmarking what's easy to benchmark - but might be totally unimportant for success.
- Impatience - skipping the planning phase to get started with the 'real' benchmarking (read partner visit), which often results in industrial tourism.

12. Aspects and Approaches to Quality

We are working to improve the performance of the manufacturer, to satisfy the customer better and create a long-term basis for business. Currently the customers are in focus for definition and development of the products that are to be used by themselves. This is one outcome from a stronger competitive market situation. The change has put pressure not only on the manufacturer, but also on the customer.

As customers, we have to be more conscious of our demands and needs and we really have to express these to the producers. We are moving from the situation where we had to select products from an assortment, and toward the situation where no final products are defined. We also see the trend of a greater differentiated consumer market where the different salesmen and stores aim at only a uniform part of the market. Stores for example offer exclusive Italian shoes for people with good financial resources, whilst the store next door offers cheap children's wear for less wealthy families.

To change the old paradigm of producing products and then let the sales people sell them, is a major task. Not because the employees do not understand the change in the market, we all understand and experience this daily. It is because the organization inside the company is reluctant to change, it is because the economic results are based on short periods, it is simply that few or none are willing to take the responsibility for the company to change dramatically even though the market has changed dramatically. If this change is to be real it is dependent on everybody in the company, everybody have responsibility for their work performance and to cooperate across functions. This requires that it is possible to cooperate, not only practically, but most important that the workers have time for systematical cooperation to improve performance. Part of our job should perhaps be 20% time to talk and work together with the internal suppliers and customers.

Currently, western countries only use the term Total Quality Management. This term consist of management *and* operational activities necessary to improve the performance. In this book we use the term Total

Quality Management (TQM). And in this chapter we are concentrating on some of the *operational* activities involved in the strategy TQM. These are Statistical Process Control (SPC), certification, and maintenance. All these activities can not be successful without support and conviction from the management. A short description of some of the views on this important issue is described at the end of this chapter.

12.1 TOWARD TOTAL QUALITY MANAGEMENT

12.1.1 Quality: Some definitions

What is quality? Today an unknown number of definitions describe all the different approaches and thoughts of the term quality. We like to think of quality as :

- systematic;
- conformity;
- zero defects;
- cost-saving;
- customer satisfaction;
- safety;
- etc.

Everybody has their opinion of what quality is, so the first assignment will be to find out what the specific customer means when he/she says that they want quality products. Then we can start the work to achieve quality. Some like to refer to their products as products with good quality; this is a wrong use of the term quality. What is good quality? Let us first try to define the terms related to the quality aspect. Here are some of the definitions taken from the ISO 9000 series, and some are defined in ISO 8402. (ISO 9000-9004: 1987):

> *Quality policy: the overall quality intentions and directions of an organization as regards quality, as formally expressed by management.*

> *Quality management: that aspect of the overall management function that determines and implements the quality policy.*

Quality system: the organizational structure, responsibilities, procedures, processes and resources for implementing quality management.

Quality control: the operational techniques and activities that are used to fulfill requirements for quality.

Quality assurance: all those planned and systematic actions necessary to provide adequate confidence that a product or service will satisfy given requirements for quality.

Product and service quality (Feigenbaum, 1983):

The total composite product and service characteristics of marketing, engineering, manufacture, and maintenance through which the product and service in use will meet the expectations of the customer.

For Total Quality Management (TQM) there exists as many definitions as there exists written books on this issue. We nevertheless understand and in this book define it like this:

Total Quality Management is the collection of all technical, administrative, creative and social activities in the entire company. TQM aims to create a basis for continuous improvements in all processes and products to satisfy all customer demands, both the specified and unspecified ones. It will lead to Total Quality which is the sign of the ideal production system where all activities are performed right the first time and all times.

12.1.2 Total Quality Management

Total Quality Management (TQM) comprises three main activities (Imai, 1986):

- maintenance of quality
- quality improvements
- quality renewal

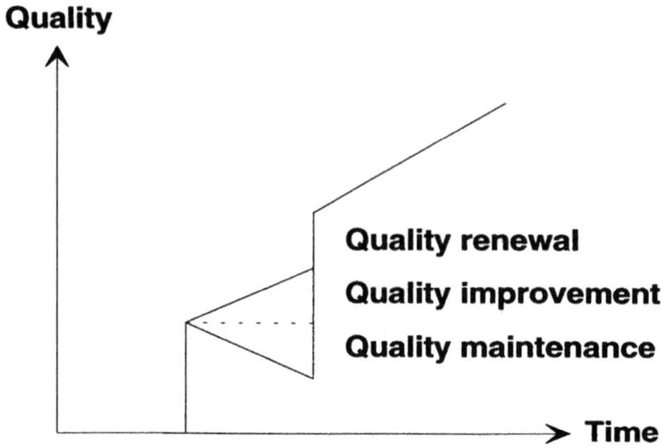

Figure 12.1 Main tasks of TQM.

All these together can make a competitive advantage through systematic work and focus on quality work, see Figure 12.1.

We recognize TQM in companies that have:

- focus on customer satisfaction for all work (internal and external customers);
- total control of all processes;
- methods for continuous improvement;
- quality as first priority;
- trained managers that understand and agree on the importance of quality.

Definitions of quality are valid for both *products* and *services*. In this book we concentrate mainly on products, but recognize that present and future products have a much more comprehensive content than the 'physical thing'. There are attributes like training, equipment description, design, financing, customer care, etc. which can be filed under services.

A very important aspect of all quality work, especially with the Japanese, has focused on *team-working*. In TQM we distinguish between two main team groups:

1. Quality circles
2. Problem-solving groups (cross-functional groups)

Quality circles are workers in the same department or working in the same function and the people volunteer for group cooperation work to solve problems that emerge. Problem-solving groups on the other hand consist of people from different functions and are working together for the purpose of solving difficult problems and trying to create a base for continuous improvements and create and make synergy of cross-functional cooperation.

12.1.3 Quality control - the early days of quality work

The industry that first was concerned with 'quality products' were the suppliers for the military. They early realized that the products had to work as specified, or they would give unpredictable results. A strong effort was done to do things right.

Acceptance and sampling tests have till now been the most common 'quality' work. No permanent improvements will be created internally in the company through this type of work. With this kind of quality work it is quite easy to see that the non-quality goods are removed, and that the quality of the remaining products fulfill the requirements for what is specified. When statistical methods are used to measure the performance, then few will deny the misuse of inside and outside specification as the only performance measure.

The need for quality control, in the sense of pre- and post-manufacturing testing is relevant for some industries. The emphasis though, must be on eliminating sources of defaults and problems when these problems have been put forward by quality control. Recent years have shown a move from emphasis on quality control to emphasis on continuous improvement.

For quality control we must successively define:

1. The purpose of the collected data.
2. Where to find the right data.
3. How to find the reliable data.
4. How to use the data right.
5. Adjust the collection process to incorporate new areas that need improvement.

The objectives of collecting data are, according to Kume (1992):

- monitoring and controlling the production process;

- analysis of the non-conformance;
- inspection.

Quality control is necessary for the above mentioned reasons. It is up to the managers to use the comprehensive data collected by the quality control activities, and to continually improve the performance. The many years of misuse of control has given the word a negative sound. Control has been linked to the performance and work done related to one person or a group of persons. Responsibility to the performance of a process have previously been linked to the operator of a machine, but now management realize that the operator has to be capable of doing the right things based on the right training with availability of tools and machinery that works as specified. If these requirements are not present then the worker is not responsible for the performance, rather the managers.

12.2 STATISTICAL PROCESS CONTROL (SPC)

The quality control concept just after the second world war was expanding rapidly because the need for large-scale production emerged. With this mass production the number of complaints increased to producers making unsatisfactory products. In the beginning this trend was moderated with a final test of the product, only removing the worst cases of products heading for the customers.

Statistical Process Control (SPC) is a collection of techniques used today by manufacturers to monitor and control the processes for fabricating products with the goal of improving the *processes*. Good processes generate good products. The prerequisite for SPC is operators and managers trained in statistical methods, and absolutely motivated for handling measuring equipment to collect data and using the statistical data. Top management has the responsibility to ensure this. A trend toward statisticians as mangers is occurring because of the need for necessary understanding and knowledge in statistics. Also school systems recognize the need for a general increase of statistical understanding in the business population. Both as a worker and as a customer our ability to understand the basic statistical way of thinking will lead to a quality focusing, valuable both for ourselves and the society in which we live and work.

Improvements in performance for the manufacturing process is to:

- remove sources of faults;
- reduce the variation of the processes.

The first is achieved through quality control, with the necessary actions taken to permanently remove the fault source. Concerning the second one, variations can be accurately found through statistical methods. But also here the necessary actions must follow the uncovering of a broad range in the process readings. Here we will show only the most central issues of statistical methods for process control.

Simplified SPC consists of four steps:

1. Define the measuring data (specifications).
2. Collection of data.
3. Processing of data.
4. Using the statistical results for implementing changes for improvement.

These four stages relate to the Deming Circle or the PDCA (Plan, Do, Check, Act) circle. When the company works through the stages of the circle, the circle will roll on toward improved performance. This is a hallmark of the company increasing its competitive advantage in the market.

12.2.1 A statistical tool in manufacturing

Recognizing major saving potentials in doing the things the right way from the start, Taguchi introduced in the 1950s the loss model for statistically supervized products, see Figure 12.2. Taguchi expresses the quality of the product as the total loss to the society by the product. The smaller the loss is, the better is the quality. Taguchi saw that the final goal for the manufacturing process was to:

- reduce the standard deviation to a minimum at all times;
- centering the nominal value.

This will minimize the loss, and maximize the quality. A very interesting and clarifying explanation.

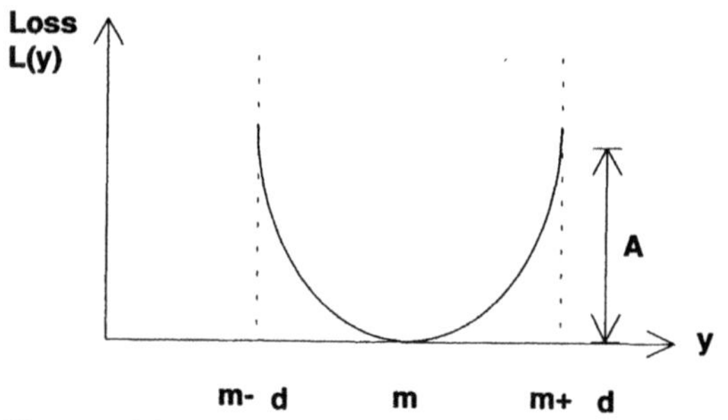

Figure 12.2 Taguchi's loss function.

Formulas for Taguchi's loss function are as follows:

A = L(y)

A	is loss in money due to the limit of tolerance
L	is the loss function
y	is real value

d = y-m

m	is nominal value
d	is deviation

$L(y) = c(y-m)^2$

$c = A/d^2$ (a constant)

The formula for the loss function shows that the loss is increasing exponentially when the performance varies for the processes. The emphasis is on controlling the variability of the process. We acknowledge that everything varies. So what do we do to reduce this variability? Variability in processes are shown through presentation of data in various diagrams. We want the measured data to lie within control limits, which in statistical control is referred to as upper and lower control limits (UCL and LCL). If these limits are not based on specifications calculated from the natural process capabilities, problems will certainly occur. We might find the process in control but out of specification. This implies that the specification must be defined by statistically trained employees. When the control limits are defined, the control of the process can be shown with Taguchi's loss function to indicate the loss the process induces. The loss is reduced through a centering of the process and reduction of the range, see Figure 12.3.

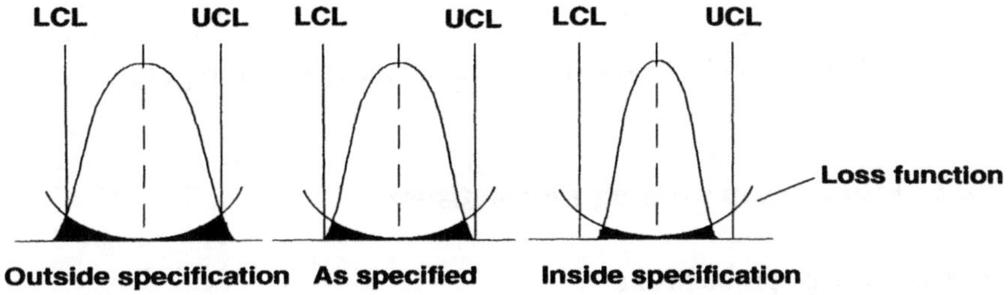

Figure 12.3 Producing outside and inside the specification induces loss (the shaded area).

Now that we have recognized variability as the problem and a key to quality control, we find that there are two types of variability (Fellers, 1992):

1. Common cause: natural variability deeply rooted in the process (situation) that cannot be removed by normal means by the local employees in the short-term.

2. Special faults: variability beyond the natural capabilities of the process (situation) that can generally be addressed by the local employees.

Variations occur in materials, measurements, machines and men. These are what Kume (1992) referees to as the 4 Ms of variation. Variations in quality characteristics appear as the sum of these four types of variation. The basic quality control activities at the factory are repeated analyzes and

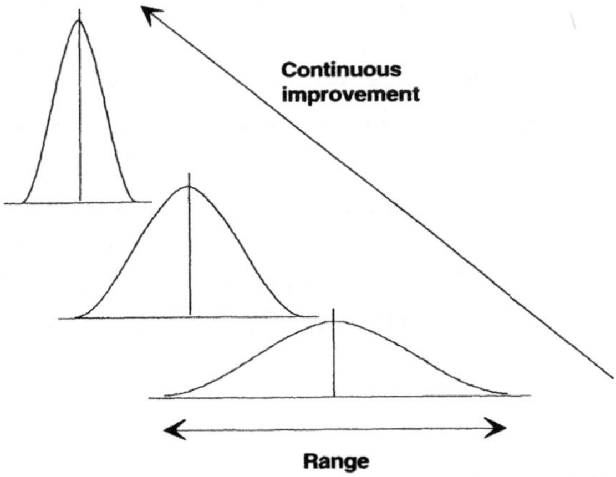

Figure 12.4 Reduction of variation (range) induces continuous improvement.

improvements in order to reduce quality variations. We see that continuous improvement is a result of reduction of variation, shown in Figure 12.4.

12.2.2 Difficulties in working with statistics

We acknowledge problems in:

- the reliability of the collected data
- the organization that is handling the data

Statistically-based decisions are reliable only when the statistical data is a result of statistical methods, and from an operator that has experience in the process. This promises a fully trained organization in statistical training and thinking, from the CEO to the engineers to the operators. With knowledge of the process and the machinery for that process, the operator will be capable of keeping the process under control and will know when adjustments must be taken.

Lack of understanding statistics is a major reason that focus on centering the processes for reductions in faults, have not had a breakthrough. Management have the responsibility to educate the employees so that they can improve the performance in processes.

12.2.3 Six Sigma

Six Sigma is best known as the strategic quality program first used by Motorola. Behind this program was a recognition of customer satisfaction as the remedy for future business. When producing complex products, the final assembled product was not a quality product even though the processes for this product were inside specification. So to produce products with zero faults, each of the manufacturing processes should produce only 3.4 p.p.m. (parts per million) defects. Without having to react to small variations in the center of the process, the process should still be within the tolerance limits of six sigma. The variation possible in such a process is shown in Figure 12.5.

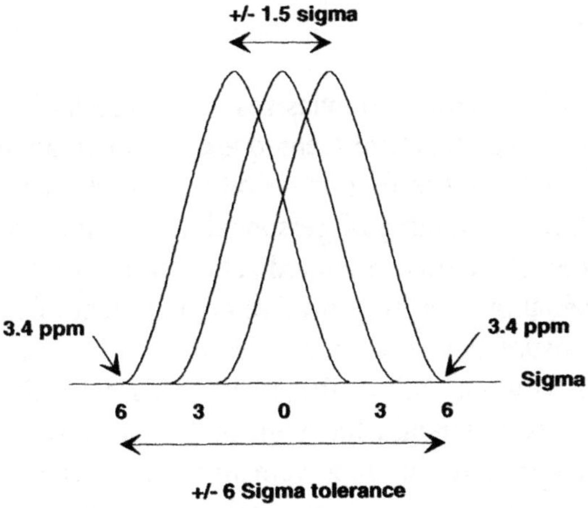

Figure 12.5 Six sigma capability.

12.3 STANDARDIZATION OF QUALITY SYSTEMS

The International Standard Organization (ISO), is responsible for the development of this quality standard. Introducing the standard we quote (ISO 9000-9004, 1989):

> *'A principal factor in the performance of an organization is the quality of its products or services. There is a world-wide trend toward more stringent customer expectations with regard to quality. Accompanying this trend has been a growing realization that continual improvements in quality are often necessary to achieve and sustain good economic performance.'*

Regarding this statement, the International Standard Organization (ISO) has developed the ISO 9000-series to ensure that customers requirements are consistently met. But this is not the objective of the certification process. The certificate is a contract document between two parties. It's focus is to ensure the demands of the customer. It has never been a tool for certification of the third party.

The ISO 9000 series is split into four subchapters, the first three (ISO 9001, ISO 9002, ISO 9003) are a collection of international standards for use on quality systems for external quality assurance purposes. The last one (ISO 9004) is for internal quality management purposes to assist in developing and evaluating the company's own quality system. In addition to this there exists several other standards, and many more are currently being developed.

The ISO 9000 series emphasizes the standardization of work performance including statistical techniques, documentation and control of processes, inspection and testing, traceability and internal quality auditing. It also emphasizes the training of personnel as essential for the success of implementing the standards. Standardization will not be achieved only through documentation, but must involve commitment of the workers. The standards will ensure that new employees involved in the work have the same guidelines to work with as the others. This helps the training of the employees and sets a standard for work performance to the workers. The standards also try to give the improvement activity a permanent result, so that the problem will not occur again. Without guidelines to the implementation of changes for improvement, the solutions will soon bring back the same shortcommings in the company.

To select the level of a desired (or needed) quality system, several factors must be considered. Building a system starting from nothing, is a very resource-demanding process. The ultimate question is whether the customers demand a specific standard for trade cooperation. The future *market* demands also have to be considered carefully by the very fact that such a quality system can not be implemented over night. We thus understand that starting to build a quality system to fit the ISO standards, a strategic plan is also necessary including top management involvement to give it the necessary importance and priority.

For companies implementing one of the ISO quality standards, this might be a way of getting the company into a system which reduces the number of incidents originally occurring due to the lack of rules for the execution of work practice. Others might find themselves in an enormous network of procedures, rules and standards which might end up actually hindering the improvement process. Long decision chains could reduce the spirit of implementing new methods and techniques. The fundamental core in the successful Japanese development in recent years, is simplification of all processes. According to the standardization of work processes, several Japanese companies find that the ISO 9000 series are actually an obstacle in their improvement process. A company must have an ISO certificate to prove the implemented quality system. For the Western companies they view the standards as a tool to make it easier to define who is responsible for the various parts of the quality process. These two contrasts show the divergence of thinking for the different cultures. This divergence is difficult to overcome when there is cooperation between the regions of the world.

To build a quality system complying with the ISO standards, the amount of direct costs has to be shown. Let us sum up some of the usual ones that are the experience of most of the firms, not saying that this is the right kind of 'result' everybody is going to find:

- internal audits (by need);
- audits by accredited firms;
- registration fee to apply for an ISO certificate;
- extra training of personnel (auditors, how to write procedures, quality awareness, etc.);
- new brochures and information material to tell customers;
- a quality staff, especially in the start-up phase;
- more and more and more paperwork.

Focusing on fulfilling a quality standard, is *not*, or should not be, the objective of the standardization work. Depending on the kind of business a company is working in, the quality demands of the customer could either be extremely high, or not so high. So, a *certified* quality system may not be necessary for most companies. What is absolutely necessary though is that there should exist a system for achieving quality, aiming at:

- handling of all quality improvement work;
- organization that prioritizes quality measures in time and money;
- documentation of the system;
- guidance and supervision so that it is meeting its objectives;
- long-term planning of all activities in the company;
- focusing on the processes for continuous improvements;
- all decisions must be made based on facts from the determined quality work;
- continuous training of all personnel;

The reason that most companies seek certification is that the customers do not have to go into the system themselves to check it. But if the certification is only to fill the book shelves then continuous improvement is misunderstood because a quality system should be the basis for continuous improvement of the performance.

12.4 MAINTENANCE IN QUALITY WORK

Definition of maintenance (BS 3811,1984):

> *The combination of all technical and associated administrative actions intended to retain an item in, or restore it to, a state in which it can perform its required function.*

Maintenance can also be split into the characteristics shown in Figure 12.6.

12.4.1 Maintenance as a part of a quality strategy

Maintenance is a part of quality assurance in the sense that repairs and maintenance service affect the product in use. If this part of the total product control fails, the term 'quality product' is not valid. Today we have a broader approach to the maintenance work, as in taking maintenance into considerations through the entire life cycle of the product, from development and design to demolition ensuring Total Quality Management (TQM) of the products.

Another approach is maintenance of the obtained quality level. This we see however, as not relevant to the ever-changing world we work in today. The most important goal for quality work is to achieve and seek ongoing improvements, and not to stabilize the quality. Maintenance work is thereby a tool for continuously improving the level of quality.

Maintenance is also about the product's life cycle, and central to this are the associated costs. Definition of the concept life cycle costs:

The total cost of ownership of an item of materiel, taking into

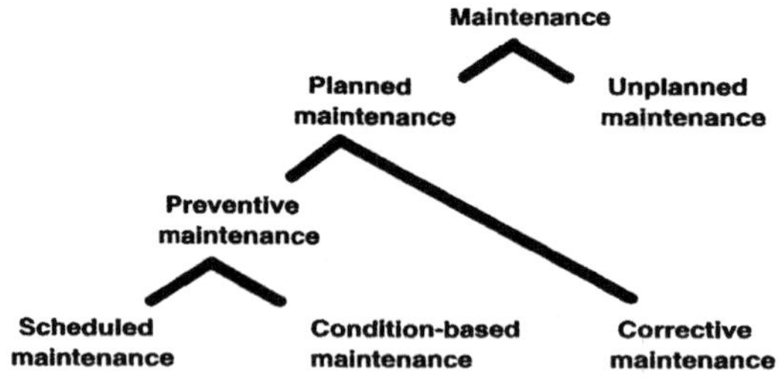

Figure 12.6 Relationships of various forms of maintenance.

account all the costs of acquisition, personnel training, operation, maintenance, modification and disposal for the purpose of making decisions on new or changed requirements and as a control mechanism in service, for existing and future items.

The main goals of maintenance are:

- efficient maintenance;
- process efficiency through maintenance;
- safety and reliability;
- quality through a life cycle perspective;

The maintenance aspect should be incorporated early in a product planning and development stage. Achieving a rational and cost effective maintenance with minimum maintenance costs is then possible, ensuring this aspect of quality is taken care of.

When performing maintenance work we must concentrate on reducing the down time and return the item to the original performance level as soon as possible. Mean Time Between Failures, MTBF, is the time from when a unit is set to work until it breaks down and is immediately replaced by a functioning part. If the unit must be repaired the expected Mean Time To Repair (MTTR), has to be considered. If the MTBF is increased and/or the MTTR is decreased the percentage down time is decreased. We must therefore:

- increase the MTBF;
- decrease the MTTR.

Maintenance has been accepted as an essential part of the corporate management structure of enterprises.

Implementation of JIT operations in manufacturing implies that only the necessary machinery should be in place and operating in the production plant. If then the bottleneck machine breaks down, it is not possible to transfer the operation to a similar machine. If the bottleneck has broken down, all the other machines must stop because there is no stock available for these machines to work on. This therefore requires no unplanned down time for all the machines if the production schedule is to be fulfilled.

12.4.2 Total Productive Maintenance (TPM)

Total Productive Maintenance (TPM) is a partnership approach of all organizational functions, but particularly between production and maintenance, for continuous improvement of product quality, operational efficiency, capacity availability, and safety. Or no downtime, no defects, and no safety problems.

The term TPM was first introduced in 1971 by the Japan Institute of Plant Engineers. Five goals were set for TPM (Nakajima 1989):

- Maximize equipment effectiveness;
- Productive maintenance system for the life cycle of the equipment;
- Implement TPM by involving all departments that plan, design, use, or maintain equipment;
- Involve all employees;
- Promote TPM through motivation management: autonomous small group activities.

Availability of the resources necessary in manufacturing is important. This is epecially crutial for bottlenecks. Measuring equipment effectiveness we can use the formula shown in Figure 12.7.

Taking into account the TPM approach together with quality control the enterprise is heading for improved products and processes.

An example of maintenance is the importance of cleaning that is strongly emphasized in quality and maintenance. Cleaning equipment can uncover problems like abrasion, looseness, dirt, etc. In a quality promoting environment a great emphasis is given to cleanness. Impurities in materials occur if the equipment is not cleaned in advance to the manufacture. Same thing happens with the work processed by operators. If their work place is not clean and clear

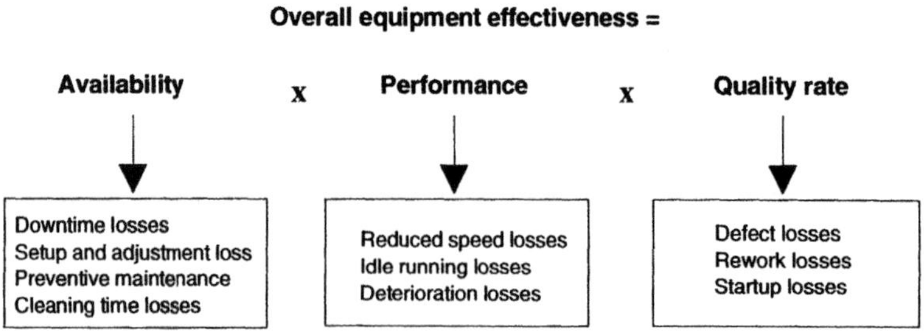

Figure 12.7 Equipment effectiveness.

we are certain that some mistakes will occur. Cleaning up the working place makes even the worker think of various improvements to ease the daily work activities.

Like TQM, top managers must incorporate TPM into the basic company policy. This includes establishment of concrete goals, training of employees, planning maintenance activities, etc.

Maintenance is an important issue for the design of the construction. When maintenance is required the operator or repairer must have easy access to the area of repairers. The recent development emphasizing quality, applies extra stress on the producers to consider the total life cycle costs. The customers are fully aware of the additional costs to products caused by abrasion and use. For producers making over-specified products, difficulties in selling the same product several times to the same customer, forces them to introduce additional, complementary, or completely new products to increase the annual turnover.

When purchasing new production equipment, the maintenance costs must be a part of the purchasing decision. When new products are developed and designed this approach must be seriously considered by the development department in cooperation with the rest of the value-increasing chain always having the end-user requirements in mind.

We have seen that total quality control involves areas like statistical methods, breaking down communication barriers, and training. But as a corporate strategy most important are the involvement of people in the quality process. The ultimate result we want is more production and a more effective organization. It is like Imai (1986) says in his book *KAIZEN - the Key to Japan's Competitive Success*:

> *Better and competitive products are the result of better people and better management, and not vice versa.*

This leads us to the next vital part of TQM.

12.5 LEADERSHIP IN MANUFACTURING COMPANIES

Managers also need to rethink how they spend their time. Our request to leadership for improvement is often met by arguments of not enough time. However, management time could be divided into three groups as suggested by Sink and Tuttle (1989), see Figure 12.8. This division in three

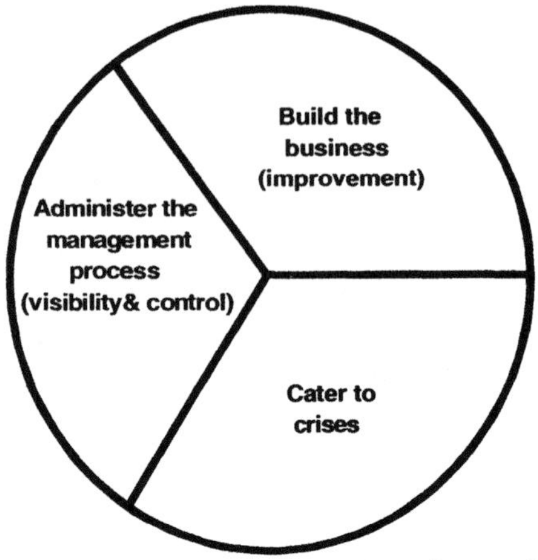

Figure 12.8 Management time dispersed over three activities.

equal sized slices is not realistic for most companies. 'Cater to crisis' is the dominating task for many managers, whereas building the business is the looser in the struggle for attention. This has turned out to be a vicious circle in many companies, because the crises are increasing due to low attention on business improvement. The Japanese management style differs radically from the Western style regarding time spent on improvement. This proves that there is little room for continuous improvement in Western companies, because most attention is given to unplanned maintenance, whereas strategic leaps through innovation are focused as the means to improve.

12.5.1 Deming's 14-point program

W.E. Deming was one of the great influences in Japanese industry's development of new and revolutionary principles of management that made a magnificent quality and productivity improvement possible. After 40 years practice world-wide he wrote the testamentary book *'Out of the Crisis'* (1986) were he advocated the necessity to change management through a 14-point program:

1. Create constancy of purpose toward improvement of product and service, with the aim to become competitive and ensure long-term survival.

2. Adopt the new philosophy, because new challenges means new tools. Management must take the challenge, must learn their responsibility, and take on leadership for change.
3. Cease dependence on inspection to achieve quality. Quality has to be built into the products in the first place. It's important to carry out inspection at the right points for minimal total cost.
4. End the practice of awarding business on the basis of direct costs. Instead minimize total cost. Move toward single supplier for any one item, on a long-term relationship of loyalty and trust. Price has no meaning without a measure of the quality being purchased.
5. Improve constantly and forever the system of production and service, to improve quality and productivity, and thus constantly decrease all costs. There must be continual improvement in test methods and ever better understanding of the customer's needs and of the way he uses or misuses a product.
6. Institute training on the job. Money spent on training, retraining, and education does not show on the balance sheet; it does not increase the tangible net worth of a company. In contrast, money spent for equipment is on the balance sheet, and increases the present net worth of the company.
7. Institute leadership. The aim of supervision should be to help people and machines and gadgets to improve. Supervision of management is in need for overhaul, as well as supervision of production workers. Leaders must know the work that they supervise.
8. Drive out fear, so that everyone may work effectively for the company. A common denominator of fear in any form, anywhere, is loss from impaired performance and padded figures.
9. Break down barriers between departments. People in research, design, sales, and production must work as a team, to foresee problems of production and in use that may be encountered with the product or service.
10. Eliminate slogans, exhortations, and targets for the work force asking for zero defects and new levels of productivity. Such exhortations only create adversial relationships, as the bulk of the causes of low quality and low productivity belong to the system and thus lie beyond the power of the work force. Posters and exhortations are often directed at the wrong people.
11. a) Eliminate work standards (quotas) on the factory floor. Substitute leadership. Rates for production are often set to accommodate the average worker. Obviously, half of the workers are above and half

below. b) Eliminate management by objective. Eliminate management by numbers, numerical goals. Substitute leadership. Internal goals set in the management without a method, are burlesque.

12. a) Remove barriers that rob the hourly worker of his right to pride of workmanship. The responsibility of supervisors must be changed from sheer numbers to quality. b) Remove barriers that rob people in management and engineering of their right to pride of workmanship.
13. Institute a vigorous program of education and self-improvement
14. Put everybody in the company to work to accomplish the transformation. The transformation is everybody's job. Any stage has a customer, the next stage. The final stage will send product or service to the ultimate customer. At every stage there will be: *Production* - change of state, input changes to output. *Continual improvements of methods* and procedures aimed at better satisfaction of the customer (user) at the next stage.

12.5.2 An overview of the major gurus in quality-promoting work

Five quality experts or gurus have gained special distinction in developing TQM strategies and operations. They are known world-wide for the importance they give in emphasizing the quality in production technology, and they are:

- Joseph M. Juran
- Edwards Deming
- Philip Crosby
- Kaoru Ishikawa
- Genichi Taguchi

Here is a short description of their work for improvements in business. Their work has been most acknowledged in the last decades after the second world war.

Joseph M. Juran

Juran started in the late 1920s to work as a quality oriented engineer. His work was successfully adopted in Japan after the second world war. He separated TQM into three: quality planning, quality steering, and quality improvements. He insists that the quality objectives should be specific to be certain of achieving quality.

W. Edwards Deming

A statistician who in the 1930s started to develop statistical methods to solve industrial problems in manufacturing. He was also working with the Japanese after the second world war, and realized the importance of statistical thinking for management to improve companies competitive advantage. The message is to identify variation in the work processes and to recognize reasons for non-uniformity. He defined the Deming circle and is also the person behind the Deming price. He created the 14 Points to help cure diseases that the management experience when changing the organization to customer satisfaction in all links in the value chain.

Philip Crosby

Crosby stresses that companies will get most out of the quality work if they concentrate on the cost of non-quality. He is the person that was first engaged in the involvement of all personnel in a quality improvement process. He defines quality as conformity with the demands or meeting the requirements. He also stresses that hard work is the only possible way of achieving improvements through quality focusing.

Kaoru Ishikawa

Japan's major developer of the Japanese TQM concept. He has been developing the 7 quality tools which have been central in quality work all over the world. His main focus is to satisfy human goals with regard to employees, customers, and owners.

Genichi Taguchi

Taguchi also stresses the statistical issues of quality. If the variations in all processes are reduced to a minimum around the mean value, then this will reduce non-conformity and reduce the costs of manufacturing best known as the Loss function. When improvements are achieved, then the mean value can be moved to a better value. Taguchi is widely known for his statistical method on parametric design. The method gives a possibility to test several factors simultaneously with few test cases, and thereby it is a relatively cheap method. The Taguchi methods require comprehensive statistical training to be successfully implemented, and this is one of the obstacles for his methods to be adopted for all companies. His statistical approaches are not implicitly acknowledged by other statisticians.

12.6 IS QUALITY FREE?

What has proved the most valuable marketing benefit for the companies adopting the quality philosophy is that 'quality is free'. To do the things right the first time should save both time and money. What we know for sure is that the cost of selling goods and services that are not good enough (not corresponding with the specifications), is impossible to measure exactly. Delayed deliveries, extra costs in telephone/correspondence, warranty costs for complaints, lost market shares due to bad reputation; all these will accumulate to a considerable amount of cost. The question is if the extra spending for eliminating faults in design, extra costs in high tolerance manufacturing machines, assurance of the functionality of the product, etc. is outweighed by extended market shares and longer-term relations to customers and vendors.

Quality costs can be divided into:

- preventive costs;
- appraisal costs;
- internal and external failure costs.

Preventive costs have been dramatically increased due to the quality focus. Implementation of quality specifications early in the design phase is a totally new situation for most companies. The time used for specifying the processes and products, and the increased use of high cost employees like engineers is moving the cost from reactive to preventive. Sometimes these costs seem very high, and these costs are more visible and tangible.

Measurement of quality costs is essential in convincing managers and owners (stockholders). Of course long-term benefits are difficult to see immediately after the implementation of a quality system, and patience is not known to be the best characteristic of profit-seeking businesses. But the necessity of quality products and processes is absorbing the entire industry in the '90s.

Why then is management the vital factor for a successful implementation of quality as the guiding star? This has to do with money, were it comes from and where it ends up. The management is always looking for profit and savings. The obstacle for a fully committed management is the lack of awareness for the advantages created by quality focusing, together with the extra money it costs to start the process. Most managers understand however, that quality and productivity work best together.

13. QUALITY AND PRODUCTIVITY IMPROVEMENT TECHNIQUES

We try here to group the techniques and tools into logical subsets, according to these headings:

- The 7 traditional tools for continuous improvement, based on statistical and graphical principles. Almost every company has distributed its own handbook on use of the tools, and the tools have proven their usefulness.
- The 7 new tools for continuous improvement, is a collection of tools that promote a more creative and effective approach to quality planning and project management.
- Techniques aimed at process analysis, improvement and waste elimination. Included in this group are techniques like process analysis and method study.
- Techniques, or rather principles, of organizational design that can help improve quality and productivity.

13.1 THE 7 TRADITIONAL TOOLS

The seven traditional tools, it seems, are in some ways not so traditional after all. There seems to be some disagreement on what seven tools belong to this group. Different sources list slightly different tools. When looking up references, you might therefore find some inconsistency between the references and those we have presented. This version of the list is based on an article series presented in Quality Progress throughout 1990:

- the flow chart;
- the cause-and-effect diagram;
- the control chart;
- the histogram;
- the check sheet;
- the Pareto chart;
- the scatter diagram.

The flow chart

It is fairly obvious that before trying to control and improve a process, one must first understand that process. Still many companies set out on improvement journeys without knowing how the processes work. Flow charts are a way of painting a picture of the process in order to understand it. Using a flow chart to gain insight into the process to be improved should always be the first step.

There are several different ways to construct a flow chart (Burr, 1990a); one can use pictures, engineering symbols or simply squares and rectangles. None of these can be defined as the right or the wrong way. The ultimate test of the suitability of the technique chosen is how well those who create and use the chart understand it.

The following elements can be included in the flow chart: the suppliers to the process and their inputs, the actual process steps, including methods, resources and environment and the customers of the process as well as the output to those customers from the process.

Flow charts can be applied to almost any process, both manufacturing and administrative processes, but the most substantial improvements often come from applying the flow chart to staff or administrative processes. One highly useful application is identifying deviations between the actual and the ideal process. Through constructing a flow chart for both processes, the two diagrams can be compared to both identify redundant and missing steps in the actual process. Figure 13.1 shows an example of a flow chart for the process of answering and directing a phone call.

The benefits from actively using flow charts are many, including the following:

- The people who work in the process understand it and are able to control it instead of being victims of it.
- When visualizing the process objectively, improvements can be easily identified.
- Employees understand how they fit into the overall process, which leads to better communications.
- Understanding who are suppliers to the process and customers of it, helps clarify who to work with and who to satisfy.

The cause-and-effect diagram

As developed by the late Kaoru Ishikawa, the cause-and-effect diagram is also known as an Ishikawa diagram or a fishbone chart, as the end product

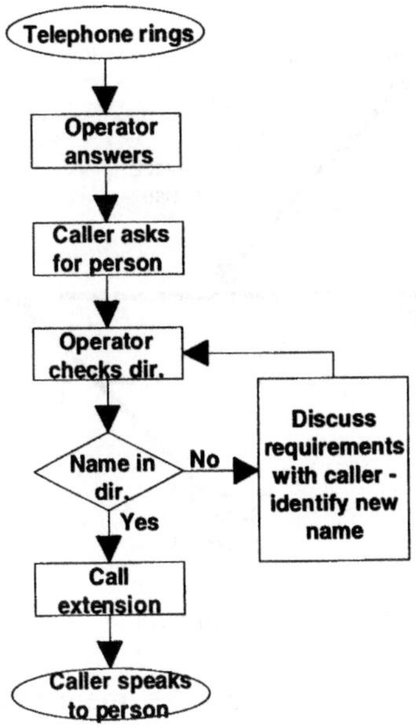

Figure 13.1 Example of a flow chart.

resembles the skeleton of a fish. The diagram is used to identify possible contributing factors (causes) within a process that influence the outcome (effect). The effect in this case is very often an undesirable outcome of the process you are trying to improve.

Before we go on to describing the actual diagram, it is worthwhile mentioning that while a cause-and-effect diagram obviously can be developed by an individual, it gives the best effect when used by a team. It is an excellent means for facilitating a brainstorming session, as it focuses the participants on the issue at hand and allows them to sort ideas into useful categories.

There are three basic types of cause-and-effect diagrams:

- Dispersion analysis, where the undesired effect is drawn on the right-hand side of a white board, at the end of an arrow. The team generates ideas as to what areas are causing the effect. These main causes/areas are drawn as branches flowing into the main branch. For each main cause, contributing causes are identified through brainstorming. The end result is a diagram as shown in Figure 13.2.
- Cause enumeration, means simply brainstorming all possible causes contributing to an effect and listing them in the order they are suggested. After the group has exhausted all suggestions, the causes are grouped

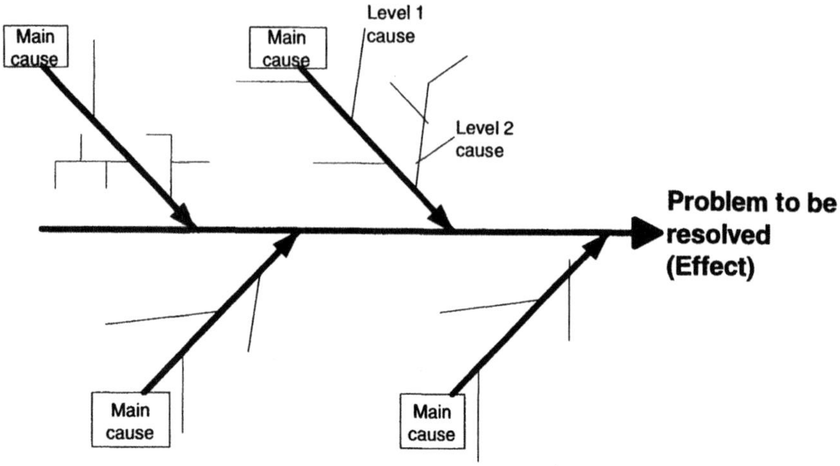

Figure 13.2 Example of a cause-and-effect diagram.

are drawn as branches flowing into the main branch. For each main cause, contributing causes are identified through brainstorming. The end result is a diagram as shown in Figure 13.2.

- Cause enumeration, means simply brainstorming all possible causes contributing to an effect and listing them in the order they are suggested. After the group has exhausted all suggestions, the causes are grouped into suitable main areas. The end result is the same as for the dispersion analysis.
- Process classification, where the flow of the process to be improved is determined by drawing up a branch containing the individual steps of the process. For each step, the group brainstorms on quality-influencing characteristics.

No matter what type of fishbone chart that is used, it is imperative that the results from the session are later used. The fish should be 'fleshed' by trying to quantify the identified causes. A Pareto diagram (which will be described later) could be used for ranking the causes. It is most important, however, that the causes identified are dealt with. Simply pointing out what the causes are, is far from enough. The causes must be attacked to be removed or improved.

The control chart

All processes display variation and for any process, there are two sources of variation; what are called special and common causes:

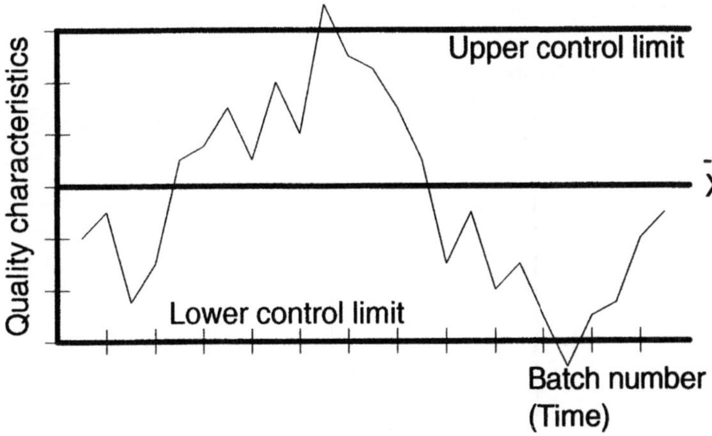

Figure 13.3 Example of a control chart.

To predict a process' behavior and to act to improve it, you will need to understand the causes of variation in the process. A means to understanding the process is to use a control chart, as shown in Figure 13.3. The control chart displays a time series or sequence of measurements of the performance of the process. Three lines are drawn on the chart: the mean value and lower and upper control limits. The control limits are calculated based on statistics to include the 3σ area in both directions.

Regularly plotting points on the chart allows for differentiation between special and common causes, and thus avoiding blaming operators for common causes. The control chart also helps to avoid adjusting a stable process, or tampering with the process, as this only increases the variation. A process is out of control if a point falls outside the control limits, if a number of consecutive points fall on the same side of the mean or if the points demonstrate a trend.

The histogram

A histogram is a graphic summary of variation in a set of data. The pictorial nature of the histogram enables us to see patterns that are difficult to see in a simple table of numbers. A histogram is shown in Figure 13.4.

A histogram displays:

- the frequency of certain events;
- distribution of these frequencies;

- variation in the process.

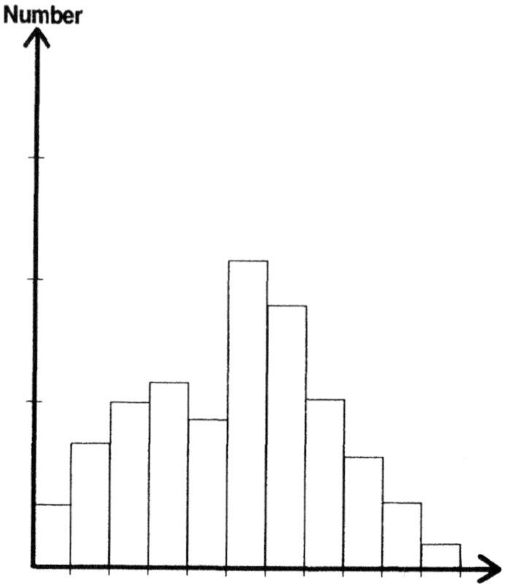

Figure 13.4 Example of a histogram.

There are some pitfalls in interpretation of histograms that are important to avoid falling into. Before drawing any conclusions from the histogram, one must be absolutely sure that the data is representative. Furthermore, no conclusions should be made based on a small sample. Finally, remember that a histogram does not show the sequence in which events took place, and can therefore not be used to show trends or stability in the process.

The check sheet

A check sheet is a means of recording data systematically about probable causes to a problem (Juran Institute, 1990). These probable causes could for instance be the main causes identified in a cause-and-effect diagram. There are no rules for constructing a check sheet, it can be drawn in whatever form is most useful for your purpose.

Figure 13.5 provides an example on the use of a check sheet, to determine with what frequency steel rods of different diameters are produced by a particular machine. As the table shoes, check sheets provide an immediate break down of data into different categories and allows for a quicker and clearer information recording than using for example a sequential log.

Date:
Machine #:
Operator:

Diameter	Check	Total
12-13		0
13-14	///	3
14-15	/	1
15-16	//	2
16-17	//	2
17-18	~~////~~ /	6
18-19	~~////~~	5
19-20	/	1

Figure 13.5 Checklist used to register the frequency of different diameters.

As with most tools, there are some interpretation pitfalls connected to check sheets too:

- Exclusion bias, giving biased results because data has not been collected from a certain place, time or under certain conditions.
- Interaction bias, where the performance of the process is affecting the process of collecting data.
- Estimation bias, where the formulas and methods use to calculate statistics from the data might introduce different types of bias.

The Pareto chart

In simplest terms, the Pareto principle suggests that most effects come from relatively few causes Burr, 1990b). Quantified, 80% of the problems come from 20% of the operators, materials or machines. More important, 80% of the non-quality costs come from 20% of the possible causes. A sound principle is to start working on these 20% of the causes, often called 'the vital few'. This is not to say that the remaining 80% of the causes should be left uncorrected; these 'useful many' should also be attended to. Working

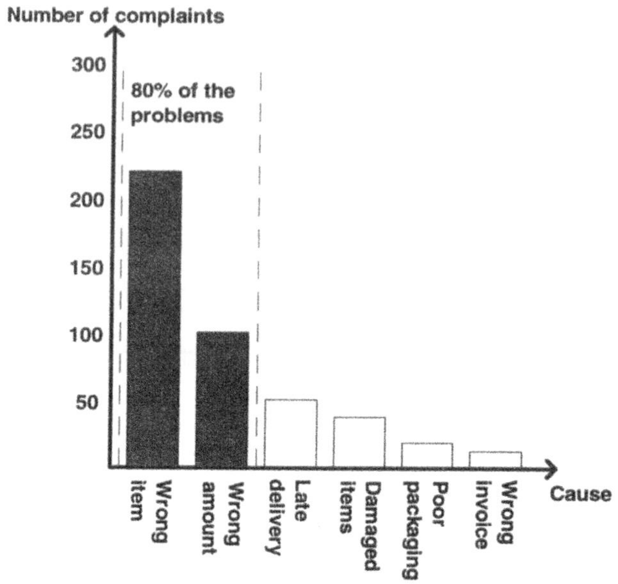

Figure 13.6 Pareto chart showing the main causes for customer complaints.

according to the Pareto principle simple says in which order to start dealing with the problems.

The Pareto chart is a form of bar chart in which the data categories are arranged in descending order of size, as shown in Figure 13.6. As can be seen from the diagram, it is easy to identify which cause the most problems so that these can be tackled first.

The scatter diagram

Scatter diagrams are the key to analyzing the relationship between two variables or process characteristics (Burr, 1990c). The data is plotted on graph paper, where the two axes represent the characteristics being examined. The resulting diagram can show many different shapes, as shown in Figure 13.7, ranging from positive correlation through no correlation to negative correlation.

It is important to use the diagram with care, however. Getting a diagram shaped as to indicate positive correlation does not necessarily mean there exists any relationship at all. Once, somebody constructed a diagram demonstrating an apparent correlation between the Dow Jones Index and the height of Lake Superior from 1925 to 1965. Likewise, a diagram indicating no correlation does not mean there is no correlation. The sample

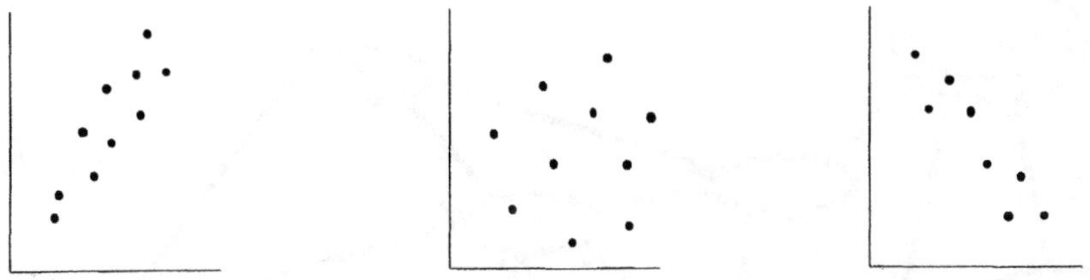

Figure 13.7 Scatter diagrams showing varying degree of correlation.

might be too small, or measurement errors or other variables might be present to blur the picture. The scatter diagram can be powerful, but must be used with care.

13.2 THE 7 NEW TOOLS

The traditional seven tools are mainly concerned with analyzing already gathered data, while the seven new tools promote how to treat problems. These are only meant to supplement the existing seven, not to replace them. Each of the seven new tools will be briefly presented below (Mizuno, 1988).

The relations diagram method

A technique to clarify intertwined causal relationships in a complex problem or situation in order to find an appropriate solution. For problems demonstrating complex networks of cause-and-effect relationships, the logical relationships between the causal factors are visualized by constructing a relations diagram. The diagram employs arrows to show the cause-and-effect relationships among a number of problems and factors that influence them (Figure 13.8). Problems are solved following this procedure:

1. Isolating all factors related to the issue.
2. Expressing these factors freely and concisely.
3. Identifying logically the cause-and-effect relationships and depicting them using arrows in a relations diagram.
4. Producing a complete picture.
5. Extracting the key factors.

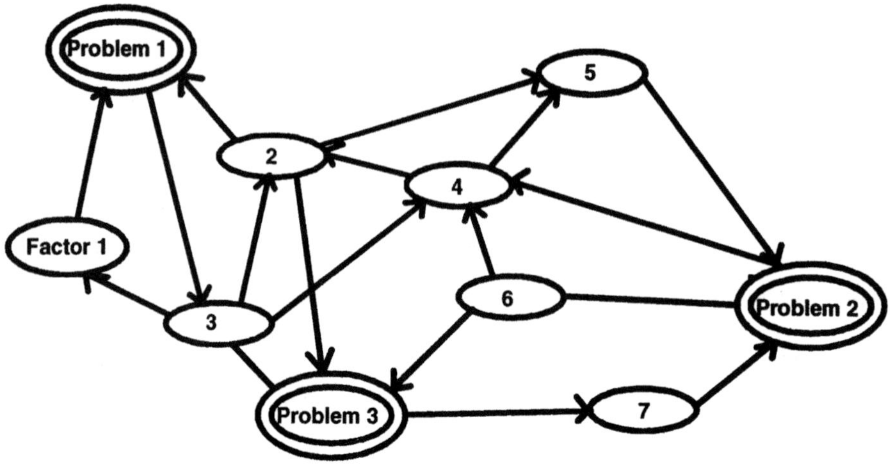

Figure 13.8 An example of a relations diagram.

Generating this diagram should lead to finding an effective solution. Repeating designing the diagram will increase the probability of finding a solution. An example of a relations diagram is shown in Figure 13.8.

The KJ method

Clarifies important but unresolved problems by collecting verbal data from disordered and confused situations and analyzing that data by mutual affinity. Based on the gathered data, a complete affinity diagram showing the relationships and similarities among the data is drawn. The method particularly concentrates on fields where there exists little or no prior knowledge or experience. The problem formulation, data collection and diagram drawing are done in groups.

The affinity diagram of the KJ method is created through the following steps:

1. Selecting a theme.
2. Collecting narrative data.
3. Transferring narrative data onto cards.
4. Sorting the cards.
5. Labeling the cards.
6. Drawing the diagram.

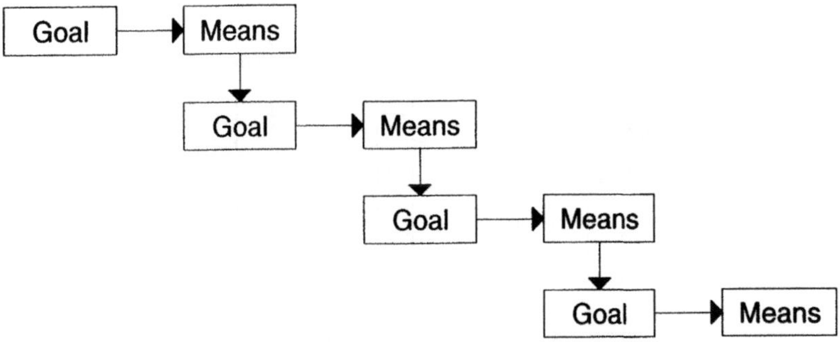

Figure 13.9 Means at one level become goals on the level below.

The systematic diagram method

For a given set of objectives, this method searches for the most appropriate and effective means of accomplishing these objectives. This is done trough the use of systematic diagrams, by which the technique clarifies important aspects of a problem, enabling an overview of the whole situation to be gained at one glance.

The systematic diagram represents events in the form of a tree and its branches. When means to achieve a goal are selected, secondary means are necessary to secure the primary means; thus the principal means become the goal of the secondary means, as shown in Figure 13.9. The systematic diagram displays the means necessary to achieve specific goals and objectives, clarifies the essence of the problem by making the subject matter visible, and searches for the most suitable means of realizing the objectives.

The matrix diagram method

The method clarifies problematic spots through multidimensional thinking by identifying corresponding elements involved in the problem and arranging them in rows and columns on a chart, Figure 13.10. The chart shows the presence or absence of relationships among collected pairs of elements. By using the intersecting points as starting points, the design allows us to both explore the problem under study from two points of view and build a base for further two-dimensional problem-solving.

Depending on the nature of the specific problem at hand, different types of matrices can be constructed, ranging from L, T, Y, X to C-matrices.

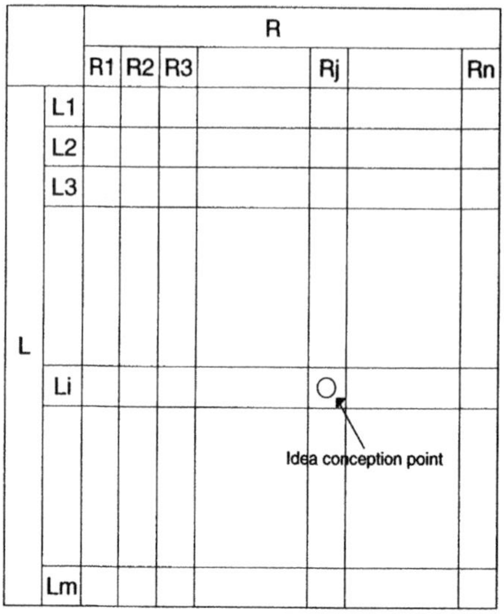

Figure 13.10 Conceptual matrix diagram.

The matrix data analysis method

Arranges data presented in a matrix diagram so that the large array of numbers can be visualized and comprehended easily. The matrix diagram arranges the items in a column/row format, with the degree of correlation entered in the relevant column using symbols or numerical values. Figure 13.11 shows an example. Looking at the matrix in the figure, it is easy to

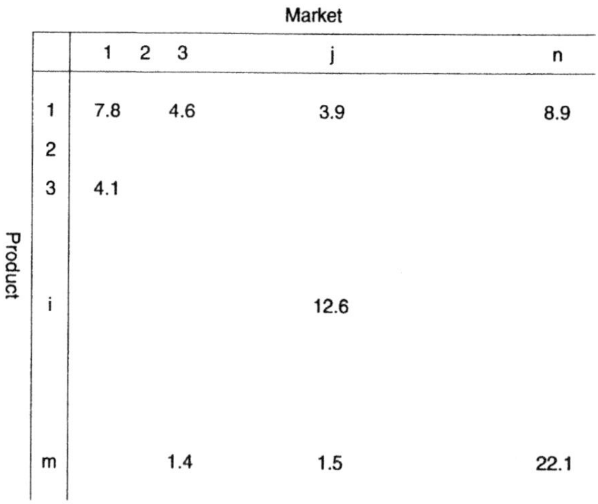

Figure 13.11 Example of a matrix analysis diagram, showing the sales of different products in various markets.

understand how helpful the matrix can be. Blank cells are naturally overlooked, while cells with symbols or numbers attract attention. Patterns of relationships can thus easily be comprehended.

The PDPC Method

The PDPC (process decision program chart) method helps determine which processes to use to obtain desired results by evaluating the progress of events and the variety of conceivable outcomes. When implementing plans, conditions change, necessitating altering the plan. This is also facilitated by the technique. By anticipating potential, undesirable outcomes of events, the technique allows process adjustments in light of actual progress. We have not provided any examples of a PDPC chart, but the method is applicable to any process, ranging from space flights to auto repairs.

The arrow diagram method

The method establishes the most suitable daily plan and monitors its progress efficiently. It is based on CPM or PERT principles and uses a network diagram connecting all elements related to plan execution, as shown in Figure 13.12. CPM or PERT rules are used to calculate optimal plans. If deviations occur, the diagram shows what other activities will be affected.

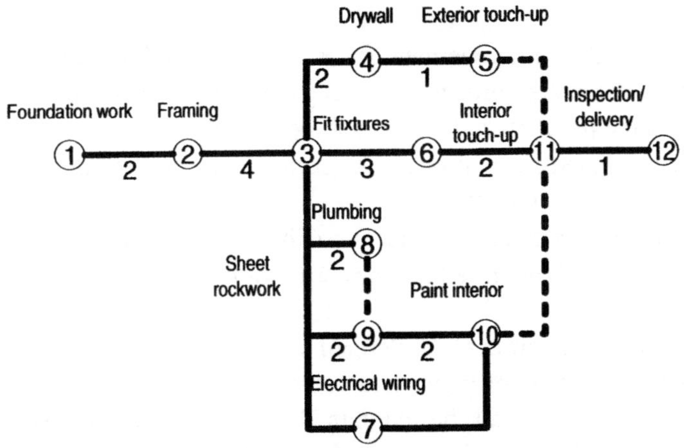

Figure 13.12 An arrow diagram for the process of building a house.

13.3 TECHNIQUES FOR PROCESS ANALYSIS AND IMPROVEMENT

The last few years, there has been a gradual increase in awareness of the importance of viewing the company as consisting of a number of horizontal business processes rather than vertical departments. Instead of making improvements within each department, that might lead to sub-optimizations, it has been realized that the processes have to improved. There are techniques to help improve the horizontal processes.

Process analysis

Process analysis is a method that provides a structured and systematic approach to managing and improving the quality and productivity of processes. Some of these processes are often cross-functional and the responsibility shared between the various departments involved. Process analysis emphasizes appointing one person responsible for the process and thus avoiding the usual disintegration of responsibility. Defining the process boundaries and its customers and suppliers allow for better communication and clearer requirement specifications.

The fundamental principles of this method have proved to be applicable to any process, including development, design and manufacturing processes (Ackerman *et al.*, 1987). Some of the principles, particular to this method, are:

- Process quality improvement focuses on the end-to-end process, i.e., includes the entire process as unit of analysis. This secures a holistic view on the problem and avoids sub-optimizations.
- Everyone manages a process at some level and is simultaneously a customer and a supplier.
- Customer needs drive the process quality improvements.
- Process simplification reduces opportunities for errors and rework.

In this context, a process is defined as a set of interrelated activities, characterized by a set of inputs and value-added tasks that produce a set of outputs. A process can be contained within, or it can span several functional organizations. Process performance refers to how effectively and efficiently a process satisfies customer requirements. A customer is anyone who is at the receiving end of materials, services or information. He can be both internal or external to the company. Supplier refers to individuals or

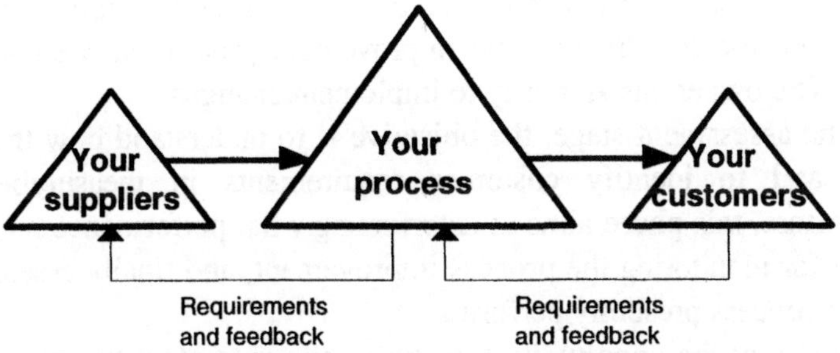

Figure 13.13 The customer/supplier model that process analysis is based on.

groups who provide inputs to the process. These can also be internal or external to the company or group. The customer/supplier model is illustrated in Figure 13.13.

The process of process analysis, as shown in Figure 13.14, is a four stage cycle, consisting of seven individual steps.

The objectives of the ownership stage are to define roles and responsibilities and ensure that everyone involved in the process

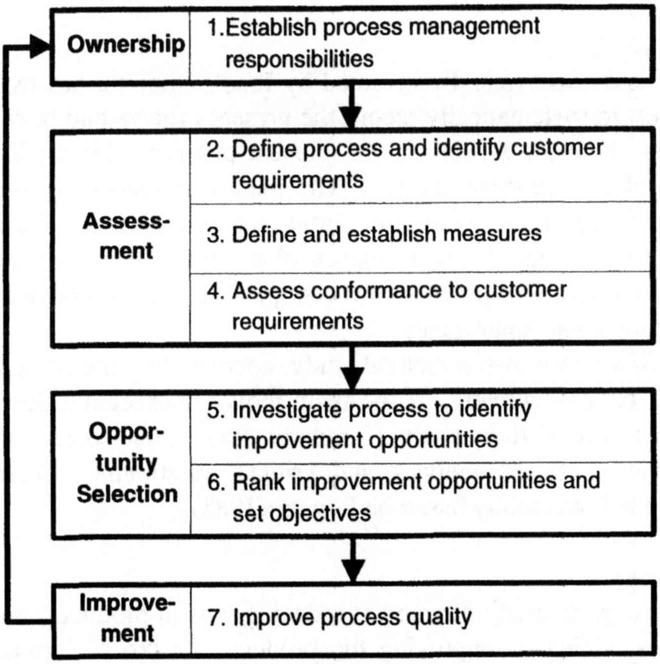

Figure 13.14 The process of process analysis.

improvement works as a team. A process owner is appointed, who is responsible for coordinating the improvement process as well as for the results. The owner has authority to implement changes.

In the assessment stage, the objective is to understand how the process works and to identify customer requirements in measurable terms. Furthermore, this phase aims at determining what performance measures to employ for monitoring the process improvement, and finally assessing how well the process presently performs.

The aim of the opportunity selection stage is to identify critical internal process problems that are affecting customer satisfaction and cost, to identify and rank process improvement opportunities and to set improvement targets.

Finally, for the improvement stage, the goal is to organize a quality improvement team to develop an action plan for how to attack the identified opportunities for process improvements. After such a plan has been developed, the root causes for the problems should be identified and removed. The process performance must be monitored and assessed on a regular basis.

As the improvement targets are reached, the targets should be recalibrated and the whole process should recycle.

Method study

Method study, as originally invented by Frank Gilbreth, has two principal objectives: to systematically record the present process and to raise critical questions about all aspects of the present process. This is, in principal, quite similar to process analysis. The main differences are that process analysis is much more focused on letting the requirements of the customer of the process guide the performance of it. Furthermore, process analysis strongly emphasizes process ownership, even for processes spanning several functional departments.

The first phase of the method study, documenting the present process, has been covered already, as the main tools for executing this phase are different forms of flow charts. Similarly, there exist some tools for use during the phase of analysis and critical questioning. These will be presented below, mainly based on Lawlor (1985).

Value analysis

This tool is particularly directed at reducing cost in products and materials while maintaining or improving the product's or process' function. It is used after the product has been put into production (Juran, 1988).

Value analysis generally proceeds through seven stages:

1. Introduction, what is to be studied, objectives with the study and who will be involved.
2. Information collection, obtaining all relevant information, e.g., usage, costs, demand, quality standards, etc.
3. Speculation, brainstorming sessions to explore all ideas.
4. Analysis, produce cost estimates of ideas and select those with the greatest promise.
5. Detailed study, each idea selected is explored in detail.
6. Recommendations, proposals are made to senior management for decision.
7. Implementation, putting to life the recommended changes.

Work simplification

This is nothing more than a formal name for a quite general and logical way of working with improvement. Groups consisting of people involved in the process to be improved are formed to go through the steps of work simplification. Having the workers themselves generate ideas, and initiate and implement the improvements are based on the assumption that most people have creative potential for improving their own jobs. Self-initiated change is perceived positively, as opposed to imposed change, which is likely to evoke fear and hostility.

Work simplification follows the following five steps:

1. Select the job/process to improve.
2. Get the facts and make a chart.
3. Challenge the facts by asking questions concerning the purpose of the activity, location of it, the sequence it requires, who does it and with what equipment.
4. Develop a better method.
5. Install the improvement.

Step 4, developing a better method, might be harder than it seems, but this step can generally be made much more effective through using some of the problem-solving techniques of the 7 traditional tools.

Critical incidents

This is a method for accomplishing, for example, step 1 of the work simplification technique, i.e., selecting a process to improve. It is a more open-ended way of seeking information on organizational problems. The term open-ended is used because the participants are completely free to express their views in any way they wish.

A selected group of employees is asked to answer questions like what incident in the last week did you feel was most difficult to handle or caused the greatest embarrassment. Recurring incidents or the most important ones, according to the Pareto principle, are selected for further improvement.

13.4 PRODUCTIVITY-PROMOTING ORGANIZATIONAL PRINCIPLES

The way the organization is designed can dramatically impact the performance of that organization. These are mainly principles from the TQM philosophy, which also support the implementation of more process-oriented management.

Interdisciplinary teams

The traditional way to organize a company has been to divide the required tasks between different departments, where each department has assumed responsibility for a set of tasks, usually highly interrelated. This is clearly seen in any traditional organization chart displaying departments like procurement, personnel, financing, marketing, production, etc. Once put into such a square, it has often seemed as if the lines of the square soon come to define rigid boundaries within which to stay. Communication across these boundaries is limited to a minimum, and a member of a department will naturally only carry out tasks that fall within the domain of his parent department. Each department seeks to maximize their influence and authority while optimizing their performance within the departmental boundaries. The result is that the whole system is far from being more than the sum of the individual elements, in fact in the worst cases, it is less. Every department sub-optimizes within the defined domain, leading to conflicting goals and actions between departments and poor overall organizational performance.

Obviously, this is a worst case scenario, but you would be surprised to realize how often it is very close to being true. As the competitive pressure of all markets has steadily risen, an awareness has emerged that this way of organizing the company's operations simply will not suffice. Every department and individual of the organization has to communicate and work toward the same, shared goals and objectives. It will no longer do to perform a mediocre operation and pass the result on to the next department with both blaming the other when the third one finally has to correct the mess. This growing awareness and its consequences have been pointed out by several different fields of expertise. Organizational experts have developed numerous interventions to move the organization in the right direction. These will be dealt with in the next part of this chapter. Psychologists have spoken warmly in favor of the more organic and human organization, and Total Quality Management has embraced these principles.

All of these areas generally advocate the use of so-called interdisciplinary teams. An interdisciplinary team can be defined as a group or team consisting of members from different functional departments and layers of the organizational structure. Such teams can be formed on an *ad hoc* basis, to investigate or solve a specific matter or problem, to be dissolved after completion of the task. Or the organization can permanently change to consist of several interdisciplinary teams, though this is more unusual. Somewhere in between these two, is the alternative of having both at the same time. This means that the organizational members mainly function within their departments, but sometimes resolve to join their appointed teams, often for a specified number of hours of the day or week.

The benefits with interdisciplinary teams are that the views and needs of different departments are present at the same time within the environment that set out to perform a task. Before making any decisions or making any physical transformations that can be hard and expensive to reverse, all views are presented. As the final outcome from the team reflects and considers the needs of every field involved, the quality of the output is probably of superior quality compared to what would have been achieved the traditional way. Even if the same outcome was reached by a functionally divided organization, the path to the output would probably be full of rework and change orders. This takes time, demonstrating another of the benefits reaped from employing teams, the lead times are considerably reduced, which is becoming increasingly important (Stalk and Hout, 1990).

Problem-solving teams

As we mentioned above, TQM is one of the areas that has advocated the de-building of the rigid functional barriers of traditional organizations. Two specific approaches to the use of teams are problem-solving teams and quality circles. The latter does not rely on an interdisciplinary composition of the groups, but problem-solving teams are a true application of interdisciplinary teams. Problem-solving teams in TQM are used as part of the company's effort to make continuous improvements, and in this context, to solve specific problems.

Before we go on to describe the teams, we will briefly repeat the two types of variation displayed by a process. A process, and this can be any process, not just manufacturing processes, will demonstrate common causes for variability. These common causes are natural variability deeply rooted in the process or situation, which cannot be removed by typical means by the local employees on a short-term basis. The removal of such inherent causes is never an easy task.

The second type of variability is local faults, with variability beyond the natural capabilities of the process or situation. These faults can be traced back to a specific and readily identifiable cause, and can generally be addresses by the local employees.

The characteristics of the two types of variation are shown below (Fellers, 1992):

Local faults

- variable out of control on a control chart;
- likely short-term solution;
- assignable cause probably readily identifiable;
- subject for morning meeting;
- solution mostly local;
- the old approach of 'putting out fires' may be reasonably successful.

System problems

- variable in control, but not good;
- total and lasting solution not local: upper management is part of the answer (problem?);
- long-term solution (one to five years is common);
- problem difficult to uncover - likely several contributing issues;
- requires interdepartmental think-tank approach;
- not a subject for morning meetings;
- will demoralize if treated as local fault;
- the old approach of crisis management will not help, something different must be tried.

System problems are seldom unidimensional, such that a single person or one functional specialist can solve the problem. In matured companies that have worked on their problems for some time, the remaining problems are usually of an interfunctional nature, and require interdepartmental problem solving. As we said before, the functional barriers between departments should be broken down, or else the result is often that the whole is less than the sum of the parts. True systems improvements often result from actions in the 'white space' on the organizational chart. The interdepartmental or interdisciplinary teams are about the only proven technique that can systematically solve system issues and at the same time improve the culture to remove some departmental boundaries.

An interdepartmental team is formed on an *ad hoc* basis, i.e., a team is formed when there is a need for one, and dissolved after the intended task is carried out. As the preceding discussion explained, the task is normally to solve a system problem, discovered either by intuition, clear sense, or statistical process control. The team should consist of about four to eight people, belonging to several different functional departments and layers of the organization. All departments involved in the problem in one way or the other must be represented, but one or more of the members should preferably come from neutral departments, i.e., departments not directly involved in the problem.

Meetings are to be held only when there is a 100% attendance, and experts can be called in as the need occurs. The meetings should be held in a 'sacred' place, meaning none of the members' home turf and a place where there are no interruptions. Furthermore, the meeting must be held during normal working hours to avoid stealing from the members' leisure time. Before formation of the team, upper management must be aware of the problem they set out to resolve and agree to follow through with all recommendations, subject to budgetary constraints.

The team normally employs some or all of the tools described earlier in this chapter. In the beginning, some theoretical discussion must be expected. These discussions may not seem to lead to any solution-related items, but will generally help to educate the members and to assess each other. As time passes, the meetings will be more productive and solution-focused, and often, the meeting frequency can be reduced without risking that the members neglect their regular work. The outcome of the team's work should be an opportunities list with cost estimates, priorities, timing and actions needed to implement the solutions.

Side effects that do occur from working in such teams are an enhancement of the company culture, development of the team members' skill and motivation and increased team spirit within the organization. All experience shows that these effects will occur, but it would be a mistake to schedule them and follow them up if they really did occur.

Quality circles

As opposed to problem-solving teams, quality circles consist of members from the same work area and layer of the organization, and they are a permanent institution. Some of the main characteristics of quality circles are that a quality circle consists of a group of persons:

- from the same work area (they have the same types of jobs and experience the same types of problems) and are under a circle leader (in the beginning, preferably the supervisor or foreman),
- who voluntarily participate in regular meetings during normal working hours for approximately an hour per week to, according to their own priorities, identify, analyze and solve problems in their own work area and give a written and oral presentation of solution proposals with cost estimates to the person(s) who have the authority to make a decision about implementation (Aune, 1985).

To enable the circle members to solve these problems, they are extensively trained in different problem identification and analysis techniques.

As with most interventions aiming at improving the organization or the organizational culture, the objectives are twofold:

1. To enable the individual worker to develop his creativity and creative forces and to utilize these abilities;
2. To strengthen the company's productivity and competitiveness.

Problem-solving teams generally require some sort of organizational changes, as these teams spend considerably more time over a shorter period of time than do the members of the quality circles. Implementing quality circles does, thus, not require the organization to make changes. Another pronounced difference between the two types of groups is that the problem-solving teams attack the vital few problems, while the quality circles set out to deal with the trivial many (classification according to Juran, 1979, who later renamed the trivial many to the useful many, Juran, 1988, 1989).

Put into a larger organizational picture, a quality circle interacts with the rest of the organization as shown in Figure 13.15 (Cummings and Huse, 1989).

As we already have said, each circle has a leader who is responsible for training the circle members and guiding the weekly meetings. Coordinating the efforts of several circles is a facilitator. The facilitator trains the circle leaders and helps them get their groups started. As the process proceeds, the facilitator may attend circle meetings to observe progress or give hints on how to improve the working of the circle. The facilitator also helps the circle to get input from experts when needed. Above the facilitators is a steering committee, who usually consists of the facilitators and representatives from the major functional departments of the company. The steering committee determines policies and procedures for the circle program and coordinates the different training programs.

Research has shown quality circles to give results in many cases, but a surprisingly large number of implementations have failed. Findings show that the failures are often due to management not giving the circles the necessary authority to implement solutions found, leading to disillusionment of the members.

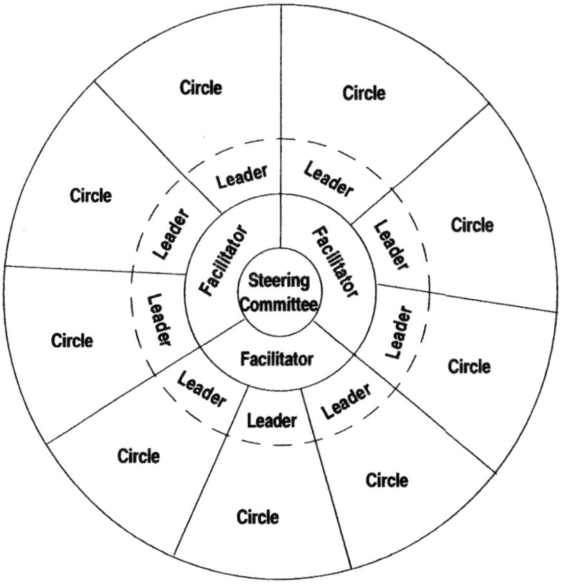

Figure 13.15 Quality circles and the rest of the organization.

PART FIVE

Performance Influencing Factors

14. MANAGEMENT CONCEPTS

During the first half of this century, mass production of standardized products was the main objective. Management focused mainly on maximizing output and optimization of direct labor and equipment costs, since they were the dominant elements of the product cost. Competitiveness was directly dependent on managing these costs. To cope with this environment, accounting systems that established objectives and closely monitored the productivity of individual workers as well as the utilization of production equipment were developed. Such systems are called Management By Objectives (MBO)

Today, international competition requires lower volume, flexible manufacturing of specialized products with extremely high quality and short life cycles, with continuously decreasing lead times. These requirements are met by modern management philosophies like:

- Just-in-Time (JIT)
- Time-Based Management (TBM)
- Total Quality Management (TQM)
- Activity-Based Management (ABM)

However, as already observed, standard costing is still the most widespread accounting system, although the industrial environment has changed dramatically in the last decade.

We will shortly elaborate and compare the mentioned managerial philosophies and conclude with a picture of the distribution in Norwegian industry. Each philosophy is analyzed with respect to a main focus, implementation process, applied tools and methods and main performance indicators.

14.1 JUST-IN-TIME (JIT)

The Japanese success during recent decades has partly been explained by the term and philosophy of 'Just-In-Time'. Several enterprises have gained huge productivity profits by adopting different techniques and strategies from the JIT tool box, with the Toyota motor works as the best known one. They have applied JIT practices since the 1960's, and Toyota is believed to be the father of the philosophy, even if similar techniques had been known earlier in the US.

Main focus

The JIT philosophy is based on the logical fact that no resources will be used until it is needed. The need is created by the market pull. When a new item is sold, a pull of products through the value chain is created. For the production department, this means that no products are manufactured until the sales organization pulls for a new product, and the sub-components and raw materials are not purchased until a need in manufacturing arises.

JIT is very concerned about waste, and the philosophy defines any kind of failure as waste. The company has to aim at a zero defect level all over the organization. Even if quality is not a main focus in JIT, quality improvement is an indispensable means to obtain decreased waste levels.

According to Chase and Aquilano (1992), human resources are also an important factor in JIT. Management must have a positive attitude towards the employee, and they have to be an integrated part of the implementation process.

Implementation process

There exists no standard approach for implementation, but usually the first step is to start with the manufacturing process and use a *kanban*-based production planning and control system. This will utilize the potentials of the pull-system (see 15.6) and emphasize the need of continuous improvement in the production process.

Tools and methods

There are several tools and methods associated with JIT. The target of JIT is to streamline the material flow in the production, but most tools can also be used in other parts of the enterprise. Here are some basic tools:

- *Kanban* is a card-based order system used to obtain a pull through the production system. When a resource is required, a card describing the need is sent backwards to the previous activity. The technique can also be applied for other functions. For example, the sales department can use cards to order from the production.

- Quality at the source means that if a quality problem occurs, the production should be stopped until the problem has been resolved. Defect parts should never be sent downstream in the process.
- Uniform plant loading is necessary for the JIT approach. To utilize the potentials of a pull through the production, the enterprise must strive for an even flow of both products and required resources.
- SMED (Single Minute Exchange of Die) is a method for reduction of setup times in the production.

Performance indicators

Reduction of waste (time, materials, etc.) and delivery precision are the main performance indicators in JIT.

14.2 TOTAL QUALITY MANAGEMENT (TQM)

The quality movement started in the first part of this century, but was not considered a manufacturing strategy until after the second world war, when quality-minded Americans went from the US to Japan to teach the Japanese about the advantages of quality. The strategy was first adopted by the Japanese, and they are still harvesting from the competitive advantage they gained. The work with quality started as Total Quality Control (TQC), but has later been named Total Quality Management (TQM) by the Western world. TQM is a complete strategy, including quality control and it focuses on management issues as organization, supplier relations, customer cooperation etc.

Main focus

TQM embraces all technical, administrative, creative and social activities in an enterprise to create a basis for continuous improvement. This should create a foundation for customer satisfaction that is the overall objective for TQM. This objective can be divided into:

- customer focus;
- process focus, to gain complete control over every process;
- long-range planning, and eliminate emphasis on short-term profits;
- institute training.

These activities include continuous employee training, helping the suppliers adopt a quality strategy, submitting all strategic planning to quality awareness and reporting quality improvements. Cost and time reduction through quality improvements must be emphasized.

Implementation process
The work with quality should be organized as a continuous process, divided into different programs. The program should cover:

- top management commitment;
- status analysis;
- training;
- organizing for improvements;
- quality measurements;
- improvement projects;
- quality awareness.

Tools and methods
TQM consists of management behavioral aspects, as well as operational methods required to meet its objective. All these main tools and methods presuppose employee training, and the methods in themselves are comprehensive and time consuming. TQM justifies the costs by promising increasing and lasting market shares through loyal customers, and increased profit. Patience is necessary to give TQM the chance to prove its potentials. The main methods/tools of TQM are:

- SPC (Statistical Process Control);
- team-working (cross-functional, quality circles);
- long-term focus on suppliers and customers by involving these in the development and manufacturing of products/services;
- work-unit analysis;
- process analysis;
- elimination of poor quality costs.

Performance indicators
The all-important performance indicator is customer satisfaction through fulfillment of all customer requirements and needs.

14.3 TIME-BASED MANAGEMENT (TBM)

The concept of time-based management was conceived by two members of the Boston Consulting Group. The search for the concept started in 1979, as many examples displayed the magnitudes of differences in cycle times between American/European companies and Japanese competitors. The concept was launched in an article in the *Harvard Business Review* in 1989, followed by the book *Competing against Time: How Time-Based Competition is Reshaping Global Markets* in 1990.

Main focus

The main focus in TBM is the reduction of time cycles. This is not to be done at the sacrifice of cost and quality, but in addition. Combining these three will provide superiority over slower competitors. The philosophy of time-based management consists almost exclusively of some assertions regarding the relationships between:

- time and business, where several proposed empirical rules show clear correlation between compressed time cycles and profitability;
- time and customers, saying that cost-based customers are a shrinking segment being outgrown by time-focused customers;
- time and innovation, proving numerous advantages from being a fast innovator;
- time and money, where improved time performance leads to higher productivity.

Besides this, TBM proposes a few guidelines on how to organize and structure operations for improved time performance.

Implementation process

Three basic steps are suggested to become a time-based organization:

1. Vision and decision, moving from awareness to commitment through assessing the current situation (both internal and external).
2. Making the change, by overcoming the resistance to change and engaging the organization in developing new solutions.
3. Sustaining the achieved improvement, thorough institutionalizing the new ways of operating.

Tools and methods

The two core concepts or methods in structuring work for time compression are:

- Organization around the main sequence, which can be defined as those activities that directly add customer value. They should be isolated and organized into natural sequences, which both streamline the operations and allow support activities to be done off-line.
- Continuous flow of work, the task of smoothing the flow through the main sequence and reducing cycle times.

These two tools can help to highlight the core activities and the connections between different parts of the company.

Performance indicators

The performance indicator is time reduction, and how fast the organization can provide the customer with the required products and services.

14.4 MANAGEMENT BY OBJECTIVES (MBO)

The origin of MBO is said to be Pierre du Pont who worked for General Motors (GM) in 1920 (Ordione, 1979). He developed a management system in which clear targets of inventory, return on investment (ROI), sales, output, and other important aspects of the business, were in place. They were based on impersonal statistical and financial measures. This system was further developed by GM in 1925 based on a concept that related a definite long-term objective to a standard expectation over a number of years. The principle was to make the standards difficult to achieve, but possible to attain.

Ordione defines MBO as a process whereby the superior and subordinate managers of an organization jointly identifies its common goals. Each individual's major area of responsibility is defined in terms of results expected, and use these measures as guides for operating the unit and assessing the contribution of each member.

MBO is elaborated here to present the old-style management approach. Despite heavy criticism MBO-systems and the extreme vertical organizational structure are still widespread in Western companies. They may not call it MBO, but the characteristics are very similar to those of MBO.

Main focus

Main focus of MBO is to establish individual job objectives, and the underlying theory is that performance will improve when individuals are clear about their job objectives.

The first step in managing is to define the objective before any resources are released or any time is spent trying to achieve these objectives. MBO is a system that begins with the hoped for outputs, and measures the quality of the effort and the effectiveness of resource utilization against these hoped for outputs (Ordione 1979).

Ordione describes MBO by the following:

- The emphasis is on outputs, not activities.
- All employees know what is expected of them and can tell immediately how well they are doing.
- They know they are responsible for results and have committed themselves to trying to achieve them.
- They feel free to make necessary decisions and take necessary actions to achieve the objectives
- At the end of the period, the manager and subordinate sit down once more and talk again. 'Here is what you said you were going to produce. How well did you do, and what are you going to do next quarter.'

Implementation process

The process includes the following phases:

1. The top manager and those key executives reporting directly to that person study the system and how it operates.
2. The top manager and subordinates set up measures of organization performance
3. Goal setting methods are then extended down through the organization to the first-line supervisory level through a series of meetings between the various organizational units and their superiors.
4. The necessary changes are made in such areas as the appraisal system, the salary and bonus procedures, and the delegation of responsibility. Ambiguous policies are clarified, and procedures that may be blocking effective operation of the system are amended. A system of 'responsibility accounting' by the cost department is also installed.

Tools and methods

Improvement should be achieved without specific tools. It is up to each individual to improve according to the goals set.

Performance indicators

MBO puts emphasis on individual and departmental productivity measures, broken down from general financial objectives and total budgets. Follow-ups are achieved through the accounting system.

14.5 ACTIVITY-BASED MANAGEMENT (ABM)

ABM, as well as many other activity-based principles (Cooper, 1988 and Brimson, 1991), are based on Activity Analysis and Activity-Based Costing (ABC). ABC was introduced in 1987 by Johnson and Kaplan (1987). They introduced this principle as a reaction upon the traditional accounting systems, which they claim had lost its relevance. ABC's objective is to distribute the overheads' costs to each product with the aid of sundry cost drivers. Traditional costing systems assume that products cause costs to be incurred whereas ABC assumes that products incur costs according to the activities that they require.

Both ABC and ABM proclaim to be innovations developed recently, but the activity concept was introduced in the 1950's by Erich Kohler (1952). He used the term 'Activity Accounting', that has many points of resemblance with ABM. The difference today is the technological advantage that has emerged through computers. A company-wide activity-based system requires data processing possibilities beyond a manual recording system.

Main focus

ABM focuses on activities and processes within the organization. Competitive advantage and continual improvement are going to be achieved through efficient performance and cost control across functions.

The philosophy is based on seven main principles, which are (Renè Mortensen, 1991):

- activity control - not resource control;
- activities driven by customers;
- activity optimization;

- elimination of non-productive activities;
- activities performed according to specifications;
- blue-collar participation;
- continuous improvement of every activity.

Implementation process
ABM is not an administrative tool, but a way to manage and inspire people to increase the added value of every process. The implementation of ABM is done in five steps (Renè Mortensen, 1991).

1. Execute thorough process analysis.
2. Manage all processes according to customer requirements and internal strategies.
3. Analyze the use of resources for each activity.
4. Establish measurement criteria for each process for daily follow-up.
5. Maintain responsibility for each activity to ensure continuous improvement.

Tools and methods
The philosophy is activity oriented. This is reflected through the tools and methods used within the philosophy:

- activity optimization through process analysis;
- elimination of nonvalue-added activities;
- activity analysis to ensure that activities are accomplished according to specifications;
- teamwork - engagement of employees responsible for each activity;
- control every activity, and improve them continuously.

Performance indicators
The efficiency of every process is the main performance indicator. Customer satisfaction is achieved through optimization of processes and activities.

14.6 ANALYSIS OF THE MANAGEMENT CONCEPTS

As an example, both JIT and TBM aim in reducing unnecessary use of time. However, the approaches differ in the means and main focus. To compare the different management approaches, we have analyzed the concepts against a number of different requirements:

- performance indicators;
- attitude towards continuous improvement;
- effect on the internal organization;
- impact on external relations;
- influence on financial management;
- level of employee involvement.

14.6.1 Performance indicators

We have used the following tripartite of measures to describe business performance:

- achievement metrics - direct metrics for business achievement;
- diagnostic metrics - indirect metrics for business achievement;
- competence metrics - capability of future business achievement

Importance of achievement performance indicators

JIT focuses highly on costs, with the main objective to reduce all kinds of waste. The value adding process should be performed with minimal resource utilization, minimizing material and overhead costs. A successful introduction of JIT production involves reducing stocks and work in progress, and thus increasing the turnover of capital.

To reduce buffer stocks and work in progress, capacity must be increased to undertake the buffer function. Investments in new production technology are also necessary to be able to reduce setup times. This means that even if the need for capital in stocks and work in progress is reduced with JIT, capital is required to increase capacity and to reduce setup times.

The objective of TQM is to gain customer loyalty by selling satisfying products/services. Thus, concerning result measures, TQM has strongest impact on market shares. Increased market share will in turn lead to higher profits.

Investments in TQM usually have a long pay-back time. The cash flow will therefore likely be negative for many years before the efforts put into TQM give results on investments.

TBM focuses on gaining competitive advantage by fast delivery, resulting in increased sales. In addition, by reducing all cycle times, overhead costs are reduced and the turnover of capital highly increased. TBM does not focus particularly on reducing material or direct labor costs, nor total assets. There is no solid evidence that time compression leads to increased market share.

MBO focuses solely on result measures, and they are assumed to be sufficient to describe an organization's performance. The main concern is direct labor cost. MBO does not put emphasis on activities, and improvements are achieved through more effective individual task performance.

ABM is highly concerned with overhead costs through the linkage with ABC. ABC's objective is to direct overhead costs to those activities consuming the costs. Cost optimization and allocation are an important part of ABM. The philosophy does not give preference to other financial measures, because they are assumed to be optimized through activity optimization.

Importance of diagnostic performance indicators

JIT is concerned with lead time and inventory turnover. The customers are supposed to receive their deliveries at the exact time they are required. Since there is an even flow of materials in the production system, a new product is continuously finished for delivery. The inventory turnover is high because buffer stocks and work in progress are minimized.

TQM is strongly concerned with quality, effectiveness, and customer satisfaction, and medium on efficiency and delivery precision. It does not focus directly on flexibility, lead time, or inventory turnover, these factors are more a result of the work invested in TQM.

The main effects from TBM's time compression are increased flexibility and inventory turnover, and reduced lead times. Medium positive effects on delivery precision, efficiency, effectiveness, and customer satisfaction should also be expected. TBM will not directly affect quality.

MBO has no apparent impact on the diagnostic measures, but efficiency and customer satisfaction are central within the philosophy of ABM through the activity approach. When every activity is performed with the highest degree of efficiency and in accord with customer requirements, all the other diagnostic measures such as delivery, lead time, etc., will be taken care of.

Importance of competence performance indicators

JIT requires investments in production technology to increase production system flexibility to compensate for the elimination of buffer stocks. To take advantage of the even flow of products, the production layout is often product oriented, limiting future changes in product specifications. The concept is also concerned with the welfare of employees. Respect for people is one of the foundation stones in the JIT implementation.

The following measures are emphasized in TQM; continuous improvement, level of training, and team participation. A medium focus is given to suggestions from employees.

TBM enhances flexibility to some degree and recommends an increased share of the work force to participate in teams, but has no major impact on the other competence measures. MBO has no apparent impact on any of the competence measures, but continuous improvements are very important within the philosophy of ABM through eliminating nonvalue-added activities and streamlining the processes.

14.6.2 Attitude towards continuous improvement

Eliminating buffers in a pull system as JIT will visualize problems the buffers were intended to hide. Stock level reduction will thus stress the need for shorter cycle times, shorter setup times and more synchronized production. This constitutes a potentially high demand for production improvement. Through continuously challenging the system variables, the JIT approach makes a good basis for continuous improvement.

TQM is continuous improvement. Its objectives are to improve both products, processes and the organization. Long-lasting competitive advantages can only be achieved by continuous improvement. The philosophy sets increasingly higher goals, and since achieving 100% quality is impossible, there will always be room for improvements.

TBM poses a clear danger that the streamlining activities become a once-only project leaving the organization with few incentives to go on compressing time. In MBO, experience is said to be a good teacher, and goals are set by using standards. This could hardly be called continuous improvement, since no systematic tools are available for the individuals responsible for goal achievement. However, continuous improvements of every process and activity are essential in ABM's streamlining and optimization process.

14.6.3 Effect on the internal organization, process vs. system view

JIT is concerned with simplification of all tasks in the enterprise. Instead of making complex system for decision making, the responsibilities are decentralized enabling decisions to be made where the problems arise. This requires a simplified interface between all functions allowing the information to flow freely.

The activities in the value chain are the frames of reference for JIT implementation. The philosophy emphasizes a demolition of borders between traditional departments.

Team work and horizontal integration are two major organizational aspects in TQM. They generate the cross-functional, cross-departmental cooperation that is necessary to facilitate continuous improvement, and to incorporate the entire company in the work towards achieving the objective set by TQM. This requires a major change, from the system view, based on separated departments, to an integration of all departments, based on the cross-functional processes as the unit of analysis.

The TBM principle of organizing and streamlining the company around the main sequence requires some organizational changes. Throughout the main sequence, one identifies what is called closed-loops that include everyone necessary to achieve desired results. The organization is divided into a number of these closed-loop teams, which are self-scheduling and empowered to make and carry out decisions. The result is an organization based on considerable degree of teamwork. Responsibilities, and the necessary competences, are moved downward in the hierarchy to allow quick, local decision that facilitates time compression. The unit of analysis is then the horizontal processes, paying less attention to system and structure.

MBO is a top-down approach in the vertical organization. Activities are of minor interest, and the philosophy represents a pure system approach to management.

A transformation to ABM will require a shift from vertical to horizontal control. Measurements and follow-up procedures have to be directed toward the horizontal flow of work, not to the vertical organizational hierarchy. Employee participation is essential, leading to extended decision-making responsibilities

14.6.4 Impact on external relations

The value system in JIT, which the enterprise is a part of, is viewed as an extension of the internal value chain. This means that the enterprise must cooperate with the customers to ensure an even demand of goods, enabling the use of pull techniques, and pursue close relations to the suppliers to ensure reliable deliveries.

TQM involves both customers and suppliers, and they have to participate actively in the development of products and quality requirements. The interfaces between the company, its customers, and suppliers are open, and characterized by information sharing and cooperation.

The TBM philosophy points out that every company should view itself as part of a supply chain competing with other alternative chains to serve the end customer. Independent time compression within a chain is not sufficient. Cooperation between the company, suppliers, and customers is required to increase the competitive advantage of the entire chain.

MBO has no influence on external relations. On the other hand, customer requirements are the driving force of every activity in ABM, and stresses the importance of the customers as a part of the planning process. The organization also has to put emphasis on their supplier relationships, because the input to the production process will greatly affect the process and its output.

14.6.5 Influence on financial management

It has been proven that improvements in quality and other operational performance measurements are not intercepted by traditional accounting systems. Coupled with the fact that TQM investments have a long pay-back time, financial management based on cost measures will not likely acknowledge the results of the TQM effort. TQM requires an accounting system based on more operational performance measures.

TBM considers time as the ultimate performance measure, superior to cost, which is seen as a lagging indicator showing up too late. Accounts represent an arbitrary allocation that could change dramatically through choosing another dividing key. Thus, TBM wants to move from the traditional accounting information and focus more on time measures.

MBO is totally integrated with the standard costing system, and they were developed together. The principle of the organizational cost centers is in accordance with the cascading of objectives down through the organizational silos.

To support ABM, the financial measurement system has to focus on activities, not departments. Overhead has to be allocated to activities through cost-drivers. Traditional budget figures do not support this requirement.

14.6.6 Expert-driven approach or employee involvement

To implement the JIT techniques, external experts are necessary. They must build the system and train both management and employees. To improve the system, however, it is essential that the workers themselves can solve problems within their own domain.

The initial phases of the TQM implementation should be assisted by experts. The longer-term aim should be to make these activities self-driven, based on thorough training of all employees.

The implementation of TBM does not require external experts, but a high level of top management commitment is necessary. A TBM implementation requires involvement from those in firsthand touch with the processes undergoing streamlining.

MBO looks upon management as a profession, and is totally top-down implemented. Implementation of the ABM philosophy requires experts in the field of cost accounting, process analysis and improvement, but it also requires employee participation to be successful. It has to be a close linkage between managers, professionals and employees to achieve the desired objectives.

14.7 ANALYSIS OF TOPP DATA

From the theoretically analysis of the philosophies (as they are ideally meant to be implemented), we conclude that, except for MBO, the philosophies to a large extent overlap. Many of the same tools are used, there is no conflict concerning organizational principles, and they are all dependent on performance measurement systems focusing on operational measures. They also complement each other, as they have some differences concerning focus, e.g. TQM on quality with an external focus, whereas JIT and TBM focus on time more internally. Furthermore, both TQM and JIT are very long-term focused, when, in contrast, TBM plans for fast results.

Combining these should then give a balanced view, without risking any clashes of interests.

Through a statistical analysis of TOPP data, it has been confirmed that implementations of TQM and JIT have not given the expected or desired results. The analysis has, however, also shown that this is not due to the philosophies' lacking ability to generate positive results, but rather due to the companies having misunderstood the philosophies. For instance, companies claiming to be TQM companies show no correlation with customer focus. This actually demonstrates the fact that most companies perceive quality as conformity to specifications, or in turn, waste, and thus cost.

The TOPP data shows that cost and time focuses yield much of the same effects, but time gives significantly better results. This can be interpreted by focusing on time and thereby cover the cost dimension. Supplementing time with quality and customer focus gives a good balance.

The main conclusion is that the best match between the different management concepts is achieved by balancing TQM, interpreted to mean customer focus, with JIT/TBM. A prerequisite for successful implementation, is to measure performance and track costs in a way that suited the concepts better than traditional cost accounting.

15. PRODUCTION MANAGEMENT

Production management is defined by Chase and Aquiliano (1992) as:

> *Management of resources required to produce the goods and services provided by an organization.*

Manufacturing has an important influence on the firms' performance. Most of the man hours and a great share of the value-adding process, take place in the production. To improve the competitiveness, the goal is to improve time, cost and quality at the same time. The mass production period is passed, and the new challenge is to produce customer-specified products at the time the customer wants them to a reasonable price.

15.1 PRODUCTION'S ROLE IN THE BUSINESS STRATEGY

Studies (Bredrup, 1994) have shown that a clear strategic plan and vision are very important to improve the production performance. An analysis of the answers from the TOPP questionnaire showed that companies which had their own strategic plans and visions for the manufacturing/assembly and the production planning and control departments, also had a great dependence on improvement of the defect rate, lead time, throughput time, adaptation time and batch sizes. Therefore, it is of great importance that the production part is well defined in the strategic plans, and that the companies have a clear vision of how they want their production in the future.

In the same study, the companies answered that they had satisfactorily strategic plans and visions for the production planning and control, but not for manufacturing and assembly. The study showed further that less than fifty percent of the companies had a written strategy both for production planning and control and for production technology.

These results prove that most companies have a large potential in developing a better production strategy. The influence on the production results and the low status level, indicate that it is possible to increase the performance by letting production objectives support the business strategy.

The question if the manufacturing organization makes a contribution to the business strategy, cannot be answered with a single yes or no. There are several levels of manufacturing strategy and differences in how these are matching with the total strategy of the firm. Some companies have no plan of what their factory will look like in the future. The other extreme is that the business strategy is built on a desired development of the production.

Hayes and Wheelwright (1985) have described four stages in the development of manufacturing's strategic role:

- Stage 1 - Minimize manufacturing's negative potential: 'internally neutral'
- Stage 2 - Achieve parity (neutrality) with competitors: 'externally neutral'
- Stage 3 - Provide credible support to the business strategy: 'internally supportive'
- Stage 4 - Pursue a manufacturing-based competitive advantage: 'externally supportive'

Stage 1 - Internally neutral

At stage one, production is regarded as a neutral issue in acquiring competitive advantages. The management views the manufacturing as a necessary evil and is only trying to minimize any negative impact that it might have. They have no strategy of what the future will look like and have no plans for their production. They avoid making changes in the manufacturing technology and are only replacing old equipment with similar types to keep the machinery in a flexible and reactive position. They do not challenge the present technology and invest in something that is new to the organization.

The organizations that are characterized by this situation, typically use external consultants to make strategic decisions about manufacturing. The reason is that the management does not have sufficient confidence in the manufacturing organization to let them handle such important questions.

Detailed measurements and control techniques are used to monitor the operational performance. No action takes place before this warning system indicates that something is wrong, and the companies usually use external experts to make the corrective actions.

This stage has a reactive point of view. The companies do not dare to change their manufacturing technology themselves, and the management does not listen to the employees that are working daily with the manufacturing problems. It is impossible to make a future-oriented strategy based on occasional recommendations from external consultants.

These firms tend to look at the manufacturing technology as a static parameter. Once the equipment is installed and is working well, there is no intent of changing it. Changes in technology only appear when absolutely required or when additional capacity is needed.

Stage 2 - Externally neutral

In contrast to the latter stage where the firms only want to be internally neutral, the firms in this stage seek neutrality to the competitors. The firms are passively following the industry practice, but is not doing anything to challenge the existing technology. Therefore, we denote the organizations in this group as externally neutral.

The operation planning horizon is now extended to include an entire business cycle. When comparing the major competitors in the marketplace, the performance of the whole business cycle is evaluated. The companies are not only viewing their own value chain, but also take into account the value system of the whole business.

To gain comparative advantage, capital investments are major issues, but the only way that manufacturing competence can contribute to this advantage, is through well-considered timing of the investments. They have no strategy telling them to invest in better equipment than the competitor. When the firms want to make improvements, they adopt technology that is well-known in the marketplace. They are externally neutral.

Stage 3 - Internally supportive

At the third stage, the management expects the manufacturing organization to make a credible and significant support to the overall competitive strategy. The manufacturing strategy is derived and dictated from the business strategy and is an important item in gaining competitive advantage.

The manufacturing organization has to translate the business strategy into objectives for manufacturing. The staff is given the authority to make decisions, but they always have to be sure that they are consistent with the corporate strategy.

The firms have to actively identify long-term developments and trends that may have an impact on the success of the manufacturing organization, but are still following the overall strategy. The firms still view the manufacturing department as an isolated part of the total organization, and such organizations are denoted as internally supportive.

The main difference between stage two and three is that firms in the latter group view technological changes as a natural response to changes in the firm's strategy. The only motivation for the firms in stage two to pursue advanced manufacturing technology, is to keep up with the their current practise. They are only doing defensive moves, while firms in stage three are actively supporting the objectives in the business strategy.

Stage 4 - Externally supportive
To reach the final stage of the manufacturing's strategic role, the firms have to base their competitive strategy to a significant degree on their manufacturing capabilities. In the long-range business plans, the manufacturing organization is expected to play a major role in securing the company's strategic objectives. Rather than being derived from its overall corporate strategy, the firm's manufacturing strategy have to be a driving force. It has to be developed in interaction with the business strategy and the other functional strategies.

The manufacturing organization has a great credibility and influence within the company and has strong interaction with other complementary functions. This makes it possible to extract the full potential from manufacturing-based opportunities. The manufacturing function is supposed to give a significant contribution to the companies' competitive advantage, for instance low cost or high quality.

The companies try to be in the leading edge with respect to manufacturing practices and technologies. They seek to acquire expertise long before their implications are fully apparent and so adjust the business strategy to support the expected developments in the manufacturing area. The manufacturing-oriented competitive advantages are 'externally' supporting the business strategy as distinct from the previous stage where the manufacturing strategy was derived from the overall strategy.

15.2 PRODUCTION CHARACTERISTICS

The production processes in different enterprises are seldom identical, and they make different demands on the production management. The production management systems required for managing craft industries, are quite different to those needed for large processing industries. The clue is to make out individual needs and specifications. One management system that fits well for one company, can be useless for another company due to other production conditions.

There are many parameters that are affecting the choice of production method. The main indicators are probably size and production process, but enterprise culture and marked requirements are important too. The production management systems do also have to change during the different stages in the product's life cycle.

Hayes and Wheelwright (1984) have described the coherence between product life cycle and the production process. In the early stages of the product life cycle, the small batches and job shop are likely to be used, but when the production volume is increasing and the product reaches a higher level of standardization, the use of line and/or continuous flow increases.

15.2.1 Forms of production

We distinguish between three different forms of production:

- piece production;
- batch production;
- flow production.

Each of these involves fundamental different forms of material flow through the enterprise. Table 15.1 indicates each production form grouped together with the corresponding product types and sales method.

Piece production

Production of one or a small number of products is often caused by special products initiated by direct customer orders. This includes both production of new products each time and repeated manufacture of same product with either known and unknown intervals. Examples of applications are production of ships and oil drilling platforms. Even if the enterprise usually delivers similar products, some engineering work is usually required.

Batch production

Production in discrete batches requires that the design and process specification of the products are known in advance. Therefore, the most

Table 15.1 Forms of production.

Production form	*Product types*	*Sales methods*
Piece production	Special products	After orders
Batch production	Special/standard products	After orders/from stock
Flow production	Standard products	From stock

proper application is production of standard products. The sales method is either after orders or from stock. If production is based on orders, customer orders are cumulated until a whole batch is covered. In the case of delivery directly from stock, the production is initiated on a basis of the stock level and forecast demand.

Flow production

Standard products with high volume like cars, electrical equipment, etc., allow continuous production over a long period. The equipment is quite specialized and is often grouped along a line. The products are sold from stock that is used to buffer the production demand. This production form is also denoted mass production that refers to large batch sizes.

Level of order-based production

The sales method influences the conditions for production. Order-based production requires that all production must be finished within the accepted delivery time. We can discern three discrete levels of order-based production:

- order-based manufacturing;
- order-based assembly;
- stock production.

If the manufacturing is based on customer orders, each product is manufactured and assembled directly for the individual customer, whilst stock production is based on a forecast demand, and the produced products are not yet sold. The order-based assembly is something between the two extremities. Subcomponents are manufactured to a buffer stock, but the assembly does not start before a customer order is placed.

15.2.2 Facility layout

The way the production plant is organized makes several constraints to the materials flow and production performance. In engineering workshops, the layout of the equipment is often insufficiently adjusted to the material flow. The pieces have to be transported in all directions in the workshop. This involves a very complex material flow. The opposite example is found in the process industry. Here, flow of materials is basis of production.

We can separate between four different facility layout formats:

- process-oriented layout;
- product-oriented layout;
- cellular production;
- fixed-position layout.

In addition, we will also discuss how flexible manufacturing equipment can improve the material flow.

All layout formats are idealized, but many manufacturing facilities present a combination of two or more of the described layouts. For example, the plant can have some core product lines while the rest of the equipment are organized in a process-based layout.

Process-oriented layout

In process-oriented layout, similar equipment and functions are grouped together. As an example, all millers are located at one bounded area, while the drilling machines are located in another separated area. This is similar approach as in the engineering workshop example and require that the parts have to be transported long distances between the requisite equipment groups to be machined.

The process-oriented layout has an advantage of being very flexible. It is easy to produce different products in small batches since all operations can be done in an arbitrary sequence. This makes it easy to introduce new products and the range of variation can be high.

The relative location of the specialized process departments has a great influence on the material flow. Therefore, the clue is to minimize interdependent movement by localizing the department with the largest demand for reciprocal transport, next to each other. To obtain the optimum layout solution, several techniques are available:

- The cut-and-try technique that has for its object simply to first figure out a working localization layout and then try to improve it.
- Systematic layout planning techniques that use flow charts to illustrate the states of dependence between the departments.
- Operational research techniques that use tools as linear programming to find an optimal solution.
- Computerized layout techniques that can be used to solve complex problems.

Product-oriented layout

The materials flow performance is realized with a product-oriented layout. This means that the equipment is arranged according to the progressive steps in the production process. To produce large quantities of a limited number of products, the use of production lines or assembly lines is preferable. The equipment is then dedicated to a continuous production on a narrow product line.

If the product range is wide, production tends to be in batches and the equipment is less specialized than in pure product lines. The production line consists of several workstations linked in a chain and some kind of material handling devices to transport the products between the stations. There are several differences among line types. Some attributes are listed below:

- material handling system (belt or roller conveyor, overhead crane);
- line configuration (U-shape, straight, branching);
- pacing or cycle time (mechanical, human);
- product mix (one product or multiple products);
- workstation characteristics (workers sit, stand, walk with the line, or ride the line);
- length of the line (few or many workers in the chain);
- width of the line (number of different work tasks to each worker).

In most production lines, the products pass the workstations in a uniform time interval. The cycle time does also define the time between successive products are finished in the end of the line. The production line balancing problem is to assign tasks to a series of workstations so that each workstation can complete the work within the given cycle time in a way that minimizes the total idle time for all workstations.

Group technology

The principle of group technology (cellular) layout is to group dissimilar machines into cells to work on products that have similar attributes, e.g., shape and processing requirements (Burbidge, 1979). The main objective is to gain benefits from a product layout approach in process-oriented layouts. Cellular production has several benefits over to ordinary process-oriented layout:

- A cell consists of a few workers who form a small work team that turns out complete units of work. This gives better human relations.

- The operators' expertise will be improved because the workers produce finished products. Reputations under threat mean quick learning.
- The quality of the products will be improved because the groups make complete products under one supervisor.
- Several production stages take place in a single cell, and fewer parts have to travel through the shop. This gives less work in progress, less material handling, and shorter throughput times.
- Each equipment does fewer jobs. This means reduced tooling and hence faster tooling changes. As a result, the setup times will be reduced.
- Group technology is a good basis for automation because it forms isolated groups of equipment.

The material flow in a traditional process-oriented production environment is often very complex. Figure 15.1 illustrates possible improvements of the material flow when reorganizing the facility layout from process-oriented layout to more use of cellular production. With group technology the material flow is simplified. The parts do not have to travel there and back to be manufactured by the requisite equipment, but can follow a straight path through the plant.

Fixed-position layout

When producing products that need to remain at one location because of bulk or weight, the manufacturing equipment has to be moved to the

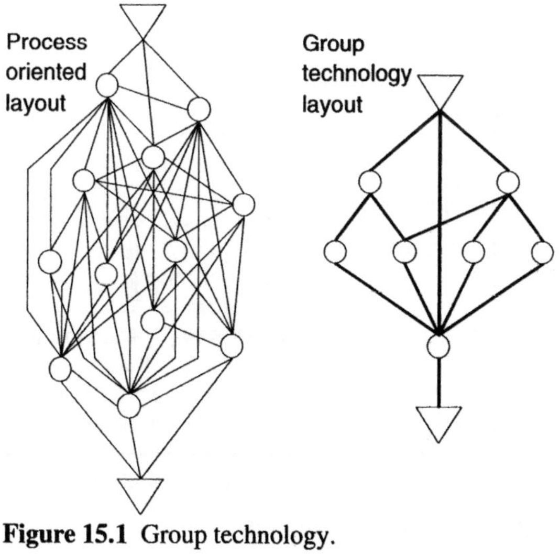

Figure 15.1 Group technology.

requisite places. In shipyards, welds often need mobile equipment. In the same way, large installations can not be moved when assembling.

Flexible manufacturing

The production equipment itself has also great influence on the material flow. Introduction of more flexible manufacturing equipment has made it possible to perform several operations in one machine. This means less material handling and a more uniform material flow.

Figure 15.2 shows manufacturing systems with different grades of flexibility according to lot sizes and product variety. The traditional manufacturing environment consists of either individual machines in a work shop or more automated production lines. Flexible manufacturing systems (FMS) are trying to combine the advantages of both kinds of equipment and can simultaneously process medium sized volumes of a variety of parts.

An FMS consists of an integrated computer-controlled system of automated material handling devices and CNC machine tools. The FMS have access to requisite product and process data and does all the scheduling within the cell. Therefore, an FMS can be treated as an independent machine doing several operations, that decreases the number of managed units and main material flows.

An FMS approach is a technical application of the group technology discussed in the last chapter. Several different operations can be performed in a single FMS cell. This means that several dissimilar products using the same cell, can be managed as a whole. Use of FMS considerably simplifies the materials management, since all the products are using the same path through the plant.

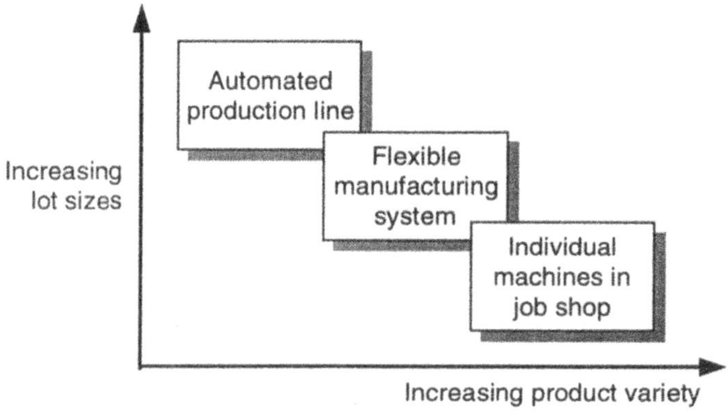

Figure 15.2 The applicability of FMS.

15.2.3 Information flow

Accurate information is a precondition to obtaining good production planning and control performance. It is important that information of what to produce is received at the right time. The information transfer can take place in two categorically different ways. Information can either be received from a centralized database as in Figure 15.3, or the information flow can go directly between the workstations as illustrated in Figure 15.4.

The introduction of computer power in production planning and control, realized the possibility of making schedules for all material needs and working progress in each workstation. This created a renaissance for centralized planning techniques and required that all information of what to produce came from a centralized database. Databased MRP systems are using this approach to solve the production planning and control problem.

The centralized approach gains several advantages. Information about products and processes are easily accessible from all involved parts, and the possibility of coordinated different production orders exists. The intention of the centralized database, was to feed the production with all necessary information and thereby improve production performance.

The centralized approach has not given such success as expected, and the method has several drawbacks. The information in the database is often not up to date and even incorrect. With wrong information about the production process, new production orders will be started according to an

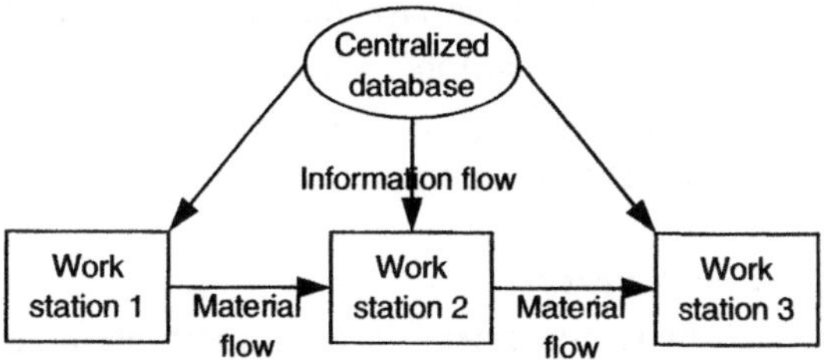

Figure 15.3 Centralized information flow.

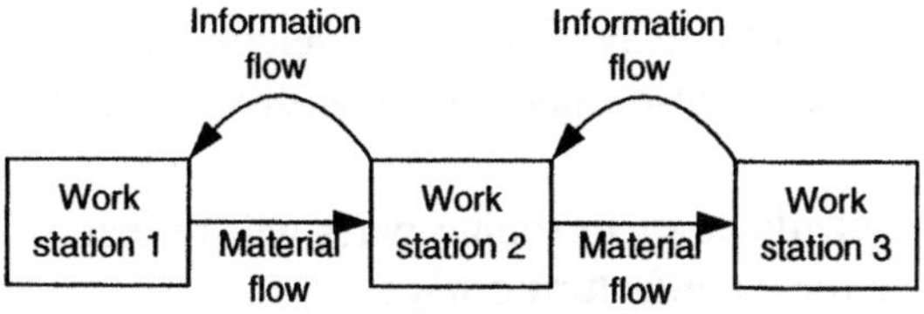

Figure 15.4 Decentralized information flow.

incorrect schedule. This means that the system will build-up stocks of some components and be short of others, and the material flow will be disturbed.

Another disadvantage is the lack of feedback from the other workstations in the production chain. Since problems are often hidden and descend from other parts of the chain, the centralized approach makes it difficult to improve the material flow performance.

In modern management concepts, there is attached great importance to cross-functional information flow. Particularly, the JIT philosophy has improved information flow along the value-adding chain as one of its milestones. Information transfer between each workstation involves a simplification of the material flow and a greater possibility of discovering potential improvements of the material flow performance. Improvements can easily be carried out because horizontal information flow permits overview and feedback from all workstations in the chain. Any unevenness in the material flow can be adjusted where the problems occur.

15.3 LOGISTICS

The production management can not be viewed as an isolated function. Porter (1985, 1990) has stressed the importance of viewing the value chain as a whole. Gaining competitive advantage requires that the firm's value chain is managed as a system rather than a collection of separate parts.

The performance of the flow of materials through the enterprise is an important factor of the competitive advantage and is a result of influence from all parts of the organization. Work in progress make a great contribution to the total working capital, but transport, inventory, and receiving and deliveries also lock up much capital.

Visualizing system weaknesses

Even if the production management is an important factor for the total results of the firm, many assumptions are made elsewhere. However, many problems in other parts of the value chain are visualized here because they do not make any difficulties before they reach the production department. To increase performance in production, problems and deviation in other parts of the enterprise are also important. Some examples are listed below:

- Poor delivery schedules of incoming parts will reflect as problems with missing parts in the manufacturing stage.

- Poor incoming quality and different tolerances from different vendors are also a matter that has its root in purchasing, but does not occur as a real problem until the manufacturing.
- Limited access to critical resources such as skilled workers or subcomponents can be critical for the production.
- Poor demand forecasts gives an incorrect master production schedule that causes several problems to the production planning and control.
- Promising more sales than the capacity limit in the production or a high level of products with low manufacturability will have a significant affect on the ability to achieve the settled delivery terms.
- Bad design for manufacturability often makes the manufacturing and assembly unnecessary complicated.
- Bad utilization of the capital equipment can be caused by poor investment due to a lack of a clear production technology strategy.

The pitfalls described, emphasizes the importance of a general view of the enterprise. Everyone has to work towards common goals that coincide with the objectives of the total enterprise.

For example, the procurement organization wants to tie up as little money in stock as possible. Therefore the level of expensive subcomponents are held at a minimum. On the other hand, the production management always need to have critical components available. Frequently, the capital demands and the most important components will clash, and there will be a divergence between basic needs.

The logistic function

To focus improvement of production and adjacent functions, the materials flow has to be the guiding principle for the organization. The top management often do not realize the importance of material flow and fails to see the potential that result from a flow-based organization.

We define material flow logistics as:

> *Management of the material flow from supplier through the production to the final customer.*

Manufacturing is an important part of the logistic chain. Even if the flow through manufacturing can be isolated, it is important to have a total view of the logistics to improve the performance of the material flow.

Often, the firms have no plain way of managing the logistics. The different elements are divided into different fields of responsibility. This make it impossible to obtain an understanding of the material flow that is optimal for the entire enterprise. All the logistic activities have to be gathered under a common umbrella to achieve optimum material flow.

Logistic-based organization

In the traditional way of organizing an enterprise, the logistic functions are spread among different departments as shown in Figure 15.5. The sales department are likely to cover the management of stocking and distribution of finished goods. Likewise, the finance department has the responsibility for the procurement because of the high money value. The production management has only influence on the receipt of raw materials and components, buffer stocks, and production planning. This involves little uniform management of the materials flow.

To improve the material flow, the logistics must be considered. It is impossible to improve the procurement or stock routines without regard to the influence at the other elements. In addition, several improvements that are hidden behind obstacles between departments, will be realized with a better coordination of the logistics function. In Figure 15.6, the logistics function is moved to its own department, taking care of all handling of materials from procurement to delivering the finished goods at the customers.

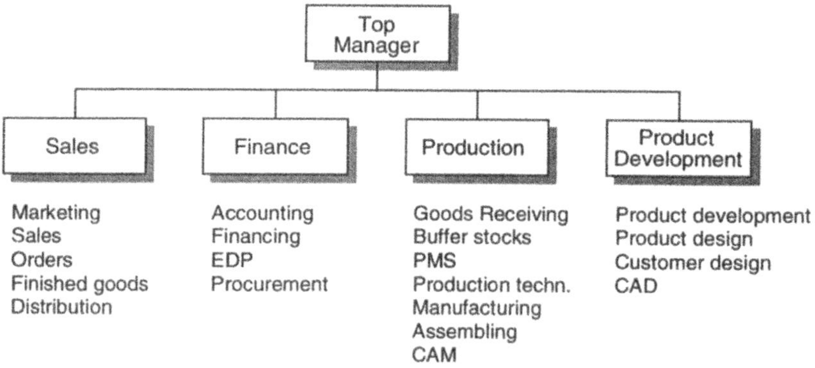

Figure 15.5 Traditional organization.

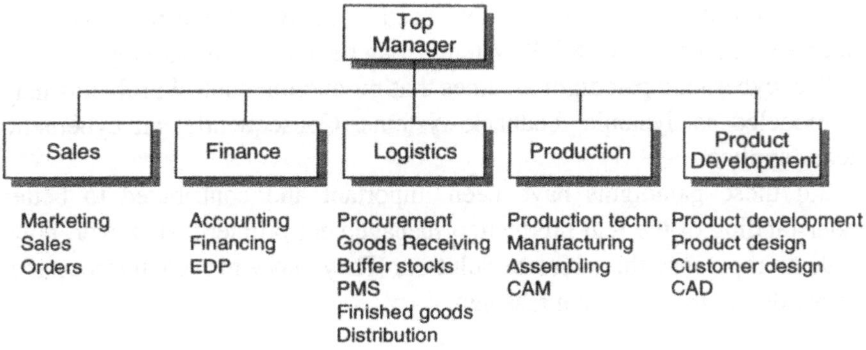

Figure 15.6 Logistics-oriented organization.

15.4 PRODUCTION MANAGEMENT PARADIGMS

Paradigms influence the way we observe and understand realities and their axioms are never questioned. A paradigm can be defined as:

> *A world belief composing the total constellation of belief, values and techniques shared by a group or society.*

Solberg (1989) introduced the term paradigm in production planning and control. According to his work, we may currently distinguish between three paradigms:

- the optimization paradigm;
- the information processing paradigm;
- the cybernetic paradigm.

The optimization paradigm is based on an axiom that all production planning and control problems can be expressed as an optimization problem. It is necessary to develop a mathematical model of the problem and then find an appropriate solution technique. This field is known as operation research and had a strong growth in interest during the 50s and 60s. Simulation belongs to this paradigm, and the latest development is knowledge-based systems.

The information processing paradigm assumes that all planning problems are mainly associated with management and processing data.

Focus is on large data volumes that need to be stored and retrieved. The success in the growth of MRP systems, may be due to this paradigm.

The cybernetic paradigm assumes that production control problems may be modeled as dynamic feedback systems. Consequently, all cybernetic theory applies.

All these paradigms have been important and contributed to better understanding of the real production management problem. However, none of them represent the ultimate solution. They have proved unsuccessful mainly due to the following reasons:

- they consider only a selected part of the total problem;
- they do not take uncertainty or stochastic behavior into account;
- they cannot handle integration aspects;
- they lack necessary flexibility to accommodate modern manufacturing technology;
- they are unable to handle continuous improvement effectively.

To challenge these problems, a fourth paradigm will probably evolve (Solberg, 1989). This may be denoted as:

- The organic paradigm

This paradigm will be based on an object oriented approach. The various units of a production system are regarded as self-contained units. For production planning and control, these self-contained systems will be referred to as autonomous planning cells. They will exchange information and cooperate efficiently and might be organized in a hetrarchy rather than a hierarchy.

The development against the organic approach is also enforced by four other factors:

- The industrial competition arena focuses on time and quality as the major competitive factors. Scheduling, risk, concurrently of activities, etc., then become important topics. This creates new requirements for production management. Information technology will play a central role.
- New manufacturing technology based on cellular manufacturing and group technology is being implemented. This requires new planning systems and decentralized systems connected in a network become

essential. The problems are connected to systems development, information management and interfacing of available software systems.

- User aspects and human/computer interaction are becoming essential.
- Modeling and simulation are going to represent important tools to design production systems.

15.5 KEY PERFORMANCE PARAMETERS OF PRODUCTION MANAGEMENT

Production management is about managing the production process. Many philosophies and techniques have been developed that include new views of the production or new technical solutions. For instance, the main target for MRP techniques is to make a time and quantity schedule for the production. Likewise, the basis of Just-In-Time production (JIT) is to manage the resources in such a way that the components occur just when they are needed.

The responsibility of production is to produce products which the customer wants at the right time and with right quality to a reasonable cost. In addition, the production must be flexible to meet any current needs and wishes and be able to adapt to changes in the future. This highlights the following factors influencing enterprise performance:

- time;
- quality;
- flexibility and adaptability;
- cost efficiency.

Delivery time

During the last years, the time factor has become an important means to gain competitive advantages. In the marked today, the customers require fast and punctual deliveries and are often willing to pay for improvement of the delivery terms. Therefore, control of the use of time in production is an important part of the production management.

The time parameter has two different dimensions:

- time to market;
- lead time.

As the products' life cycles are decreasing, the time spent from an idea to a finish product gets more important to minimize. The time to market is mainly a product development dimension, but the interaction with the product management is important to achieve an efficient and effective product development process. Concurrent engineering is dealing with this kind of problem.

The lead time is an important part of the delivery terms, and the production management has a main share of the responsibility of controlling this parameter. A product's lead time is defined as:

> *Time from when an order on a product is placed until the finished product is delivered and installed at the customer.*

The production management has mainly to different means to reduce the products lead time:

- use of stock;
- decrease the throughput time.

The easiest way is to produce to stock. If the products are stocked, the lead time will only be the time spent to deliver from stock. Even if this approach is an efficient method, it has several drawbacks. If the demand forecasts get wrong and products get out of stock, it can often last a long time until the customers gets their products. In addition, the capital cost of keeping a stock on hand is very high. On this basis, the production management has to balance the capital cost with the cost of possible delivery delays.

Stocking finished products are an extreme use of stock to decrease the lead time. The use of subcomponent storage in the production is a more cost-efficient method to reduce the drawback of high capital costs whilst still improving the delivery terms. By assembly after customer orders, a greater product flexibility can be combined with a shorter time of delivery.

The other main means of reducing the product lead time, is to decrease the throughput time in the production. Throughput time could be defined as:

> *Time from when the manufacturing starts until the whole batch is ready to be delivered.*

For production based on stocks, this means the time from when the raw material is taken from the storage until the last unit has arrived at the storage for finished products. The only proper way to solve the lead time problems, is to challenge the throughput time.

The throughput time can be divided into several isolated parts, see Figure 15.7:

- Production time is the time the batch is being served by the machine and includes operation time and the time for rigging up and down.
- Control time is the time that is used for inspection of materials and products.
- Transport time is the time used for pure transport between machines or between machines and storage.
- Total waiting time between the machines consists of storage time and waiting time before and after the machines.

The production management has influence on most of the components. Even if the operation time in the machines is often determined by the machinery technology, several techniques can be used to decrease the setup time. The best known method is named SMED (Single-digit Minute Exchange of Die). The method separates internal and external setups. Internal setups assume production stops, but external setups can be done simultaneously with the production. The clue is to transfer the setup time from internal to external time, and after that, try to decrease the remaining part of the internal setup time before working with the external time.

The control time is often moderate and is little affected by the production management. On the other hand, the transport time is very much

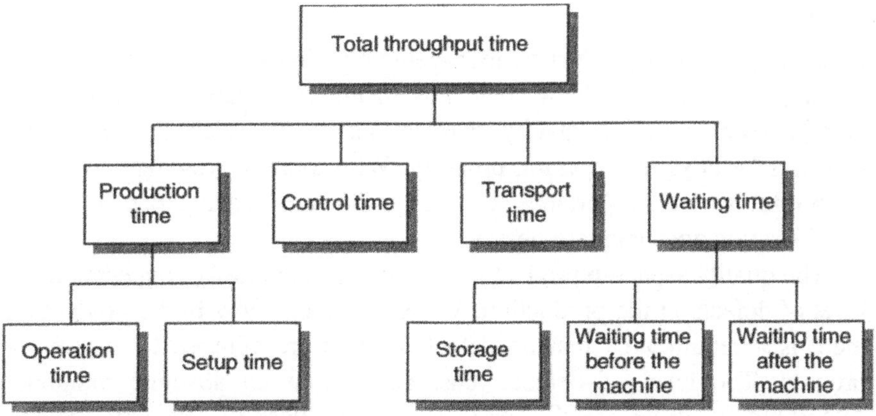

Figure 15.7 The coherence between the components of the throughput time.

affected by the production layout. The traditional process-oriented layout involves a lot of transport between the machines. Use of transport can be reduced by thinking of the production flow when arranging the equipment.

The waiting time is without doubt the largest part of the total throughput time in most enterprises, and can be decreased considerably by managing the production in more detail.

The batch sizes are a suitable means for decreasing the throughput time and the level of work in progress. This also involves a reduction of the lead time and the capital needed. In addition, a decrease of batch sizes will also increase the flexibility of the production management. Small batch sizes will not tie up the production plans for a long time, and changes of the plans can be adopted more rapidly. On the other hand, the decrease of batch sizes involves an increase of setup costs and time spent on setups.

To obtain the advantage of reducing batch sizes, setup time must be corresponding decreased. The setup time can be reduced by investing in new equipment with improved setup performance. Another approach is to improve the setup time of the existing equipment.

The batch sizes are very dependent on the forms of production. In true job production, the batch size is equivalent to the order size, but in regular mass production, the batch size can be very large, and the same product can be produced for long time.

From a logistics point of view, the best batch size is equal to the order size. If the customer wants to order unique products, this equates to one-of-a-kind production. However, this approach makes great demands on the setup time. Besides, the throughput time of the unique manufacturing process must equal to or be less than the accepted lead time.

Product quality

In creating modern production management philosophies, quality has been an important factor. New demands from the markets involve that enterprises have to take quality more serious. A customer expects that the products he buys are working properly. With today's growing complexity of products, the importance of the production management's quality consideration does increase heavily.

The quality aspect in production management is mostly connected to the level of defects in the production. To get a smooth flow in the production, there is a requirement that no defective parts are allowed to disturb the process. The level of defects must be held at an absolute minimum. According to Aune (1990), Norwegian and most of Western industry produce defects that are equivalent to about 10% of the enterprises' total turnover. In several companies, this loss is even as much as 20%. With a greater effort to decrease this defect production, Japanese enterprises have shown that it is possible to reach a level below 3-4%.

There are many approaches to decreasing the number of defects. An example is Statistical Process Control (SPC). By checking that the manufacturing process is under control all the time, any disturbance to the process can be discovered at an early moment in time, and requisite improvements can be made.

The SPC method does not improve the production management system itself. It is only an instrument to detect if there is any failure with the system. Therefore, it is important to follow up with improvements. This can be improvements of the equipment, new sequence of produced parts that require a safer adjustment with reference to the defect rate, more skilled operators, etc.

Flexibility and adaptability

The constantly decreasing life cycle of the products forces new requirements on production flexibility. Modern production management systems must handle rapid changes in the production environment. An increased number of the production orders is customer specific and therefore non-standard. The customers also want greater freedom in changing the number of items to order, and the requirements to the lead time are constantly tightening.

In the case of production management, flexibility has two different meanings:

- Production ability to satisfy changes of the customers request for products including changes in the delivery time, numbers of items and changes of the product;
- Production systems future possibility to adapt changes in the production environment.

The ability to adapt changes from the customer is closely related to the lead time of the product. If the product has a short lead time, the production orders can be frozen at a later stage. Before this moment in time, the customers can then change their orders within the capacity constraints of any limited resources.

In production systems that are based on a long-term horizon, changes in customer requests must be served by use of buffer stocks. If the requested items are then stocked, the customers' requests can be fulfilled within a short time horizon. On the other hand, use of stock as flexibility buffers has several drawbacks as earlier argued. If the requested item is out of stock and a new batch is not started or even planned, the lead time of the item will be rather long. At least, the lead time will be as long as the throughput time, but this depends on whether all raw materials and resources are available.

If customer orders are cancelled after the production of the items has started, it is often difficult to cancel the production orders. This requires that the items will be added to the stock and the capital cost will increase. Besides, if the cancelled customer orders included non-standard items, the manufactured items will not be in demand.

It is far easier if the production process has a short throughput time. This involves more flexibility to change the production in short-term. Within the throughput time, the customers can change their orders without use of buffer stocks. Therefore, a shortening of the throughput time involves decreasing stock levels without less customer flexibility.

The production systems flexibility with respect to changes in the future production environment, has a great influence on the future competitiveness of the production organization. As described previously, it is of great importance to coordinate the production strategy with the overall business strategy to adapt to the future environment.

Efficiency and cost

Ultimately, the price that the customers pay for their goods and services, has a great impact on the competitiveness of the company. The cost is highly related to the production productivity. Other parameters mentioned cover the effectiveness of production, but efficiency is important too. Proper utilization of the production equipment, efficient material flow and efficient routines and techniques, form important means to lower the costs.

To be cost-efficient, it is not enough to obtain an efficient production. The resource utilization is also important. We define resources as:

> *The machinery, equipment, people, etc., needed to support the production process.*

During recent decades there has been even more limitations of the resources. In many industries, the equipment needed to sustain competitive advantage are become very expensive. Also, on some occasions, there is shortage of some kind of resources like skilled workers and power.

When the supply of a resource makes a constraint on production, the managing of the resource becomes an important issue, and good exploitation makes for a competitive advantage. This dictates that in most industries, the limitation of resource restrictions is an important part of the production management work.

The production management systems differ in the way they react to limitation of the resources. Some companies solve possible resource conflicts on an *ad-hoc* basis, while others take the resource limit into account when making the production schedules.

However, cost and price factors are far from the only measure to use for evaluation of production. Operational performance must be used in addition to financial indicators to evaluate performance of production organizations. In the longer term, cost reductions of how well the effectiveness and adaptability parameters are taken care of in the past. It is important that the company secure its competitive advantages for the future.

Lean production strategies and requirements

Improvement of the performance parameters described, also coincide with the thoughts behind lean production. The Lean Production concept was introduced by the research team involved in IMUP (International Motor Vehicle Program) research program (Womack *et al.*, 1990). Lean production is announced as a substitution of mass production and as an answer on the increasing demand for customization of products. Very simplified, lean production wants to combine the advantages of craft and mass production to become flexible with low costs. The customer is in focus.

The main challenge today is to produce what has already been sold. Time then appears to be the major competitive factor with two dimensions, response time and life cycle and therefore evolution. Continuous improvement and the change from logic of load towards logic of flow, are the new rules of the game.

Bullinger *et al.* (1993) says that lean production is a production system characterized by improved coordination and integration of organization and technology. However, he emphasizes that lean production is a state, and not a final concept.

The ultimate goal is to improve time, cost and quality at the same time. According to Bullinger *et al.* (1993), the main strategies for a lean enterprise are:

- decentralized responsibility;

- integration of functions;
- coordinated logistics;
- process-oriented material flow;
- decentralized information management;
- integrated quality management;
- design for rational and productive manufacturing;
- manageable production structures.

These strategies are not all new and indicate sources for waste reduction. They are adapted in lean production but are central for all productivity work.

15.6 PRODUCTION PLANNING AND CONTROL CONCEPTS

Several concepts for production planning and control have been developed during the last decade. Even if they are all based on a limited number of basic principles, they attach importance to different key performance parameters and use different approaches to the planning and control problem. The introduction of computer science and solution methods based on operational research and artificial intelligence has allowed the development of more complex systems. However, there have also been developed several techniques based on a simplification of the production planning and control routines.

The most common production planning and control systems are:

- Materials Requirement Planning (MRP)
- Just-in-Time Production (JIT)
- Periodic Batch Control (PBC)
- Optimized Production Technology (OPT)
- Load Oriented Production
- Simulation.

15.6.1 Basic production planning and control principles

There is mainly two different classes of production planning principles:

- push of production orders;

- pull of production orders.

Push of orders

The traditional production planning and control techniques are based on a principle that push the orders through the production system. This means that starting of new orders is initiated by a calculated production plan. The basic illustration of a push system is that new orders are put in front of the value chain and then actions proceed successively through the operation stations as shown in Figure 15.8.

In its basic form, the push approach is a clean order-based system. This means that no action in any part of the value chain takes place before an order is initiated. Then the order is divided into demands of raw material and sub-components to be purchased and manufactured. It proceeds with the next operation in the value chain at a scheduled time when all the other requisite factor inputs are ready.

There are very few enterprises where a clean order-based model can be used without modifications. When the order point is placed in the very beginning of the value chain, the lead time usually will be extremely long. The lead time will be equivalent to the throughput time, and since no components can proceed with the next operation unless all factor inputs are ready, a deviation of only one sub-component can make a critical delay for the entire order. In addition, the average lead time will be large, which will result in a high level of uncertainty.

Unless the value chain in production consists of very few links or a very long lead time is allowed, a total order-based production is not of current interest. A common solution is to forecast the future demand, and base the production on this prediction. The time horizon of the demand forecast depends closely upon the length of the throughput time. The fact is that forecasts based on long-term forecasts involve great uncertainty. This mean that longer and more uncertain throughput times will involve greater possibility for the demanded forecasts to go wrong.

To neutralize the uncertainty of the demand forecast, the use of stocks to smooth the demand is a common approach. When the demand forecasts are underestimated, the products are supplied from the stock.

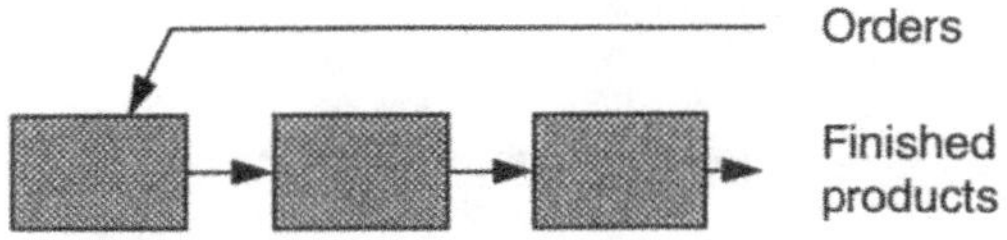

Figure 15.8 Push of orders.

Because the use of stocks is very capital demanding, not all products and subcomponents can be stocked at every moment. The stock level must be a compromise between the cost of capital and the cost of going out of stock. Therefore, the use of stock can not completely eliminate the uncertainty. There will always be a possibility of going out of stock.

Pull of orders
The other kind of scheduling technique is based on a pull of orders. This means that every link in the production chain orders the requisite factor inputs themselves. In spite of the push approach, where the order was put to the first link in the value chain, the pull approach involve that the products are ordered from the last link. Then the links have to make an order to the previous links for the needed components. This is illustrated in Figure 15.9. The result is that no parts are being produced before a need from the next link is indicated, and the products are pulled through the value chain.

Without use of any stock, the pull system will behave exactly like the push systems except for a longer and more complicated ordering procedure. When a new order arrives at the last link, new suborders are sent backwards in the chain. All previous operations must be done in advance before the final product can be made. With no kind of buffer stocks, the lead time will be the same as with the push system because the throughput time is the same. Therefore, a pull system is less suitable for order-based production than a push system, because of the work and time loss that is a result of the calculating and sending new suborders backwards in the chain.

In practical terms there must be buffer stocks at each ordering point between the operations in the chain. This will make each ordering system as an independent system. The links have to maintain their buffer stocks themselves, and the demand from the preceding link in the chain can be satisfied from stock. This results in the lead time of the production only will consist of the throughput time of the last link in the chain. All the buffer stocks will be maintained by a pull through the chain.

If the production is not uniform, e.g., if the capacity of different

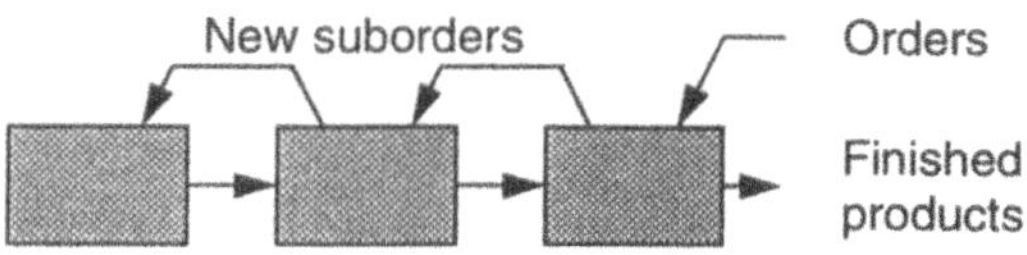

Figure 15.9 Pull of orders.

operation stations have a fluctuating utilization during a time period, the need for buffer stocks will fluctuate too. This will mean that the pull system becomes difficult to manage because of the possibility to go out of stock during peak times and build-up of stocks during slack times.

Another drawback is the possibility for the spin-off effects when fluctuating utilization occurs. If one link gets a higher demand for their components, it automatically decreases its stock level to meet future risen demands. This demands that the link has to order even more from its previous link and the cycle recurs.

To make up for these drawbacks, additional information about the fluctuation of the total demands can be passed on to the links of current interest. This means that every link will know about the future change in demand and can make some preventative effort. The information flow is illustrated in Figure 15.10.

Comparison of push and pull systems

The push and pull systems have different fields of activity. We have already ascertain that the pull system is not good for order-based production. The pull system's best mode of operation, is the case of stable production. Many unequal products can be produced, but the demand for each product must be uniform. This means that the forecast demand does not have to be known, because the pull system will adjust itself. However, the demand must not fluctuate much over time.

A major difference between push and pull systems is that a pull system does not need a long-term production plan. Orders are satisfied by buffer stocks that in turn release a pull backwards in the chain. In a push system the demand of finished products must be known in advance for a time equivalent to the throughput time of the whole system. To satisfy a shorter lead time, companies using push systems with long throughput time must base their production on forecast demands that could be unreliable.

An advantage of pull systems is that every link in the chain can be

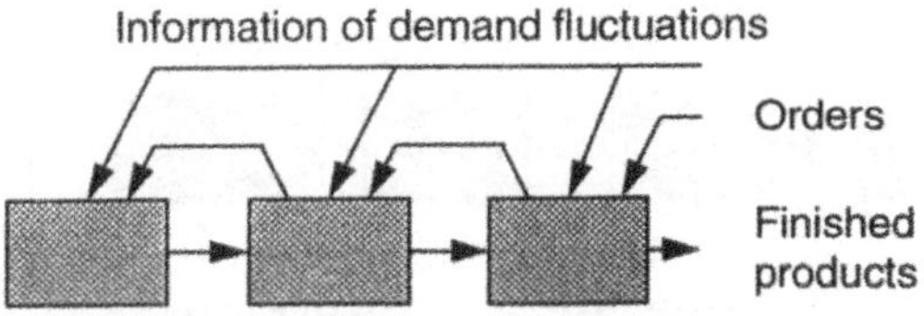

Figure 15.10 Pull of orders with additional information flow.

managed independently of each other. This approach gives better overview, and it is easier to improve the system. Each link control their own stock levels. Potential improvements are easier to discover, and improvements are often cheap and can be done with localized effort.

Most production management systems consist of elements from both the push and pull technique. As an extreme, the traditional MRP system is typically based on the push technique, but several less important inputs are often managed independently. Likewise, the production in the independent links in a push system are managed as a push system.

Push and pull systems can also be combined in a broader sense. To build an order system based of only the pull technique is very difficult, because buffer stocks are necessary to manage short lead times. A practical solution is to divide the process into operations that are independent and that are dependent of specific orders. The latter operations can be managed by a push system, while independent operations can use a pull systems to supply the push chain as illustrated in Figure 15.11.

This combination will get advantages from both approaches. Even if we assume no buffer stock in the push system, the total lead time will be reduced to the throughput time of the dependent operations. Because the demand for components used as factor inputs in the push chain is often quite uniform, the supporting pull systems will work under good conditions.

Inventory management

Both push and pull systems use stock to regulate demand. The difference is

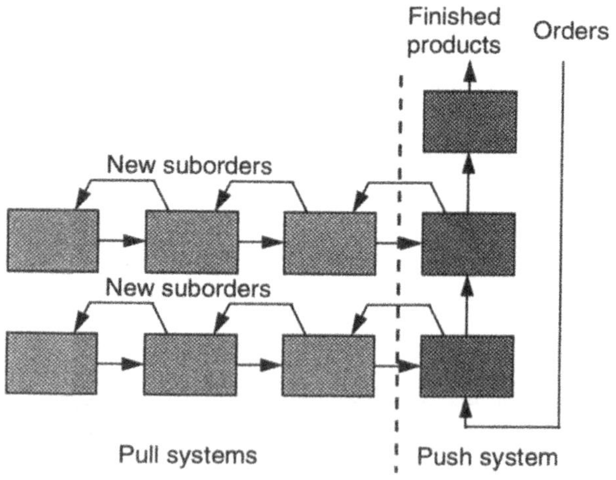

Figure 15.11 Combined push and pull systems.

that push systems use stock to neutralize the impact of production not harmonizing with current demand. When the production does not match the demand directly, the use of stock is necessary. The stock is treated as a safety stock to be used in case of underestimating of the demand. In the pull systems, inventory management is actively used to manage the production flow. Each link in the chain is responsible for fulfilling the demands and thereby maintaining the buffer stocks.

There are two different kinds of inventory control systems for probabilistic demand: the re-order point system and periodic re-order system. Each time the stock level decreases under a given level, the re-order point system orders a fixed batch size. With the periodic re-order system, orders are made with fixed intervals and varying order sizes to satisfy the given stock level.

The re-order point system responds much quicker at change in demand. The safety stock level can be very low because it is only the lead time of the ordered items that represent the uncertain period. For the periodic re-order system, any changes in the demand between re-ordering times will not be intercepted. Therefore, the safety stock level has to be higher, but as opposed to the re-order point system, the periodic system does not have to be responsive to the stock levels all the time.

Since both inventory models have advantages, no one can be said to be best. In push systems the inventories are often controlled by central management, and it is often difficult to get a total overview of the stock levels all the time. In these cases, the periodic re-order system is likely to be used. In pull systems the inventory systems can be managed independently. Each link in the pull chain has better overview of the stock levels and the re-order point approach is likely to be used because it requires less safety stock.

Bottleneck-based production

In practical production system, there can often be a small number of bottleneck resources. By strict control of these bottlenecks, the production planning and control process can be considerably simplified. Bottleneck-based production means that the schedule for the whole production process is based on the utilization of the bottlenecks. All other resources are treated as supplier and customers with adequate overcapacity. This approach reduces the production planning and control problem to that of the bottleneck. In front of the bottleneck, the production orders are pulled to satisfy the demand in the bottleneck. When the products are finished in the bottleneck, they are simply pushed throughout the rest of the value chain.

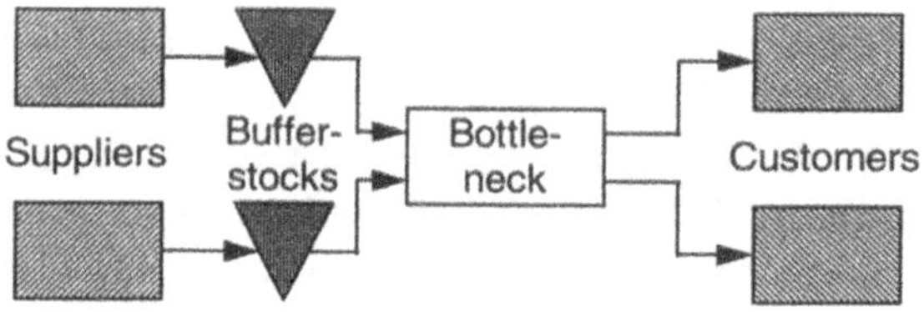

Figure 15.12 Bottleneck-based production.

Figure 15.12 illustrates a bottleneck system. In front of the bottleneck, buffer stocks can be added to balance the workload to secure a high level of utilization of the bottlenecks. If the bottleneck is an absolute limitation of the production system, it is important that the bottleneck is never out of work because this means lost utilization for the whole production process.

15.6.2 Materials Requirement Planning (MRP)

In consequence of the possibility of using large computing power in the production planning and control, the material requirement planning technique was developed in the US in the sixties, and is probably the most common production planning system today. Even though newer techniques have been developed in the resent years, most of the systems installed today are based on the MRP technique.

The original MRP technique described by Orlicky (1975), did only plan the supply of required materials. Today's systems also include links to other functions such as master production scheduling, capacity requirement planning, purchasing, and production control systems. The extended version is called MRP II, and the letters stand for Manufacturing Resource Planning. Since the concept includes the follow-up aspect and the control loop is closed, the modern systems are often referred to as 'MRP II - Closed loop'.

The MRP method uses the push system and is based on a long-term master production schedule that can include both current and forecasted demand. First, the demand for finished products is expressed in terms of net demand for components according to the product structure given by the bill of materials (parts list) for the current product. The net demands are then adjusted according to the current stock levels of the components. To satisfy the gross demand, fixed batch sizes are scheduled according to the given lead time for each component. The recalculation of demand for the

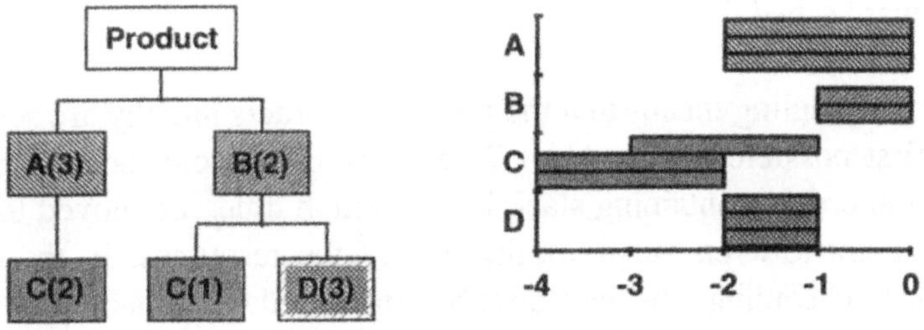

Figure 15.13 Example of a product structure.

components and their lead times is done step by step at each level until every part that belongs to the final product is scheduled.

An example of a product structure is graphically represented in Figure 15.13. Each box represents a component with can be both parts and assemblies, and the numbers in the parenthesis are defining the number of components that correspond to the item above.

MRP II also includes capacity planning. Even if the average production capacity exceeds the scheduled work during a given time period, local lack of capacity can occur after initial production planning as shown in Figure 15.14. This means that some production orders must be adjusted to obtain balance between the required and the current capacity. The scheduled work must not exceed the capacity roof in one time period.

There is principally two different approaches of adjusting the production plan to the capacity of the resource:

- forward scheduling;

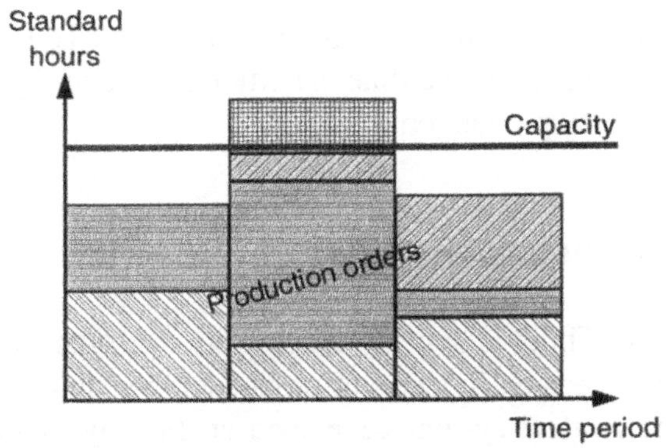

Figure 15.14 Example of scheduled workload for a work center.

- backward scheduling.

Forward scheduling means that the production orders initially are scheduled to the first possible starting day. When lack of capacity occurs, possible production orders containing slack in production time, are moved to a later starting point to even out the utilization of the resources. In the case of backward scheduling, the last possible starting day is made the starting point for the initial production plan. To adjust to the capacity requirements, production orders can be moved to an earlier starting point.

There is also provision for splitting the production lots to adjust to the capacity requirements. When lack of capacity occurs, only the requisite number of parts will be produced, and the rest will be moved to a time period when the pressure of the resources are lower. This approach adds great flexibility to the capacity planning that can adjust the capacity with less use of buffer stocks.

Introduction of the MRP technique is a result of the need to use computer power to give a more accurate scheduling of quantity and lead time for complex products. The algorithm does not challenge the length of lead times, but uses fixed lead times for each component to calculate the mean lead time for the entire product. The batch sizes and the setup times are also treated as given. This means that the system is not able to benefit from any potential improvements. The MRP system is regarded as static and simply accepts the environment within it is placed (Bauer *et al.*, 1991).

Another criticism of the MRP approach is that it require a very long-termed planning horizon. All components are planned with a fixed throughput time, and the system pushes the production orders through the production. This involves a very long throughput time for the finished product, and the system must use buffer stocks to intercept any fluctuations in the demand after the dispatch of the first operation. To be able to accept orders within a reasonable lead time, the MRP production systems must use a high grade of stock production.

15.6.3 Just-in-Time production (JIT)

The term 'Just-in-Time' is actually a complete management concept to be used all over a company. In terms of production planning and control, it is particularly the *kanban* system developed at the Toyota motor works in Japan that relates to the philosophy.

Chase and Aquilano (1992) describe the elimination of waste and the respect for people as the two fundamental concepts for the success of Japanese production management systems. Fujio Cho from Toyota Motor Company have defined the waste as:

> *Anything other than the minimum amount of equipment, materials, parts, and workers (working time) which are absolutely essential to production.*

That involves no surplus, no safety stock, no defect production. This means that nothing is produced unless it is demanded by the final marked.

Elimination of waste means that a production system must utilize every potential improvement. Therefore, as opposed to MRP, the Just-in-Time approach is based on the pull technique, which means that it is easier to detect possible improvements. The consequences of defective parts and long setup times are immediately discerned, and the system makes the stocks visible. This make a good basis for improvement.

The Japanese solution of making pull in production, is by the use of *kanban* that is the Japanese word meaning 'card'. When a demand for a component occurs, the operation station make an order by sending the former operation station in the production chain a card describing the required components. This approach secures that only components that are demanded, are being produced. Enterprises that have applied Just in Time production on a large scale also control their purchasing, transport and inventory management by use of *kanbans*, as shown in Figure 15.15.

The traditional *kanban* production only pull for single components. That means that it has to be a known demand for a specific component before a *kanban* can be send. The pull system must use buffers between the links in the pull chain. Therefore, the *kanban* system does not work well for companies that produces a lot of variants in small batches. There has to be a uniform demand for the pulled components.

Introduction of a Just-in-Time production system is not sufficient to get free improvements, but it is a tool for making the problems visible. To

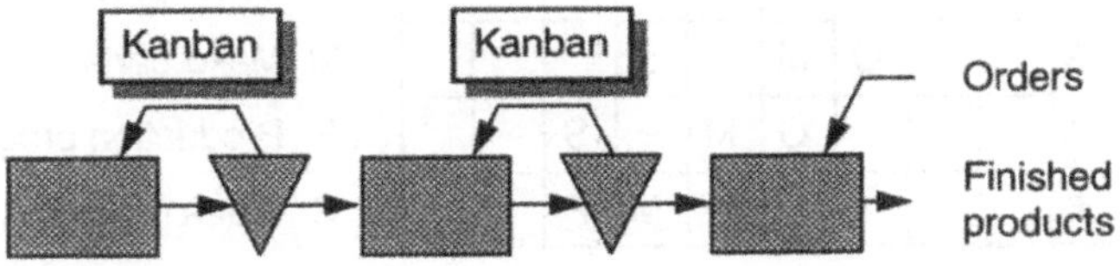

Figure 15.15 *Kanban.*

succeed with *kanban* production, there must be an associated importance given to quality and throughput time. To improve the possibility to change production after the current demand, the setup times must be reduced too. An introduction of *kanban* will probably not give the desired results if improvements are not simultaneously being made in other related subjects.

The Just-in-Time production system brings time, quality, and quantity factors into focus and creates a good environment for improvement. Flexibility must be satisfied through buffer stocks or overcapacity.

15.6.4 Periodic Batch Control (PBC)

Periodic Batch Control (PBC) was developed in England in the 1930s by R.J. Girgli, but in the present time, Burbidge has worked further with the development and takes advantage of the group technology (Rolstadås, 1988). The system only produces the amount of components during a time period that is actually demanded in the next period by the next link in the production chain. As opposed to MRP and *kanban*-based system, the PBC system uses fixed order cycles, but the batch sizes fluctuate to absorb the variation in demand.

The logic in the PBC approach is simple and a production plan based on the technique is illustrated in Figure 15.16. In the example, the order cycle is fixed to one week. The planned sale must be known at the program meeting four weeks in advance. The throughput time of the order processing and the manufacturing is fixed to one cycle period. The system also gives one slack cycle period to determine the production program because it is impossible to know the exact demand and to smooth out the production.

In the PBC system, no components are dispatched before a demand is

	1	2	3	4	5	6	7	8	9
WEEK 1	• O	M	P	S					
2		• O	M	P	S				
3			• O	M	P	S			
4				• O	M	P	S		
5					• O	M	P	S	

- • Program meeting
- O Order processing
- M Make parts
- P Production program
- S Sales program

Figure 15.16 Periodic Batch Control (PBC).

calculated. It emphasizes production of the right quantity for the right time period. The time and quality factors are treated as fixed, and the system does not offer any means to improve them. Even if the system has a possibility to tune the production in the short-term, the system requires a long-term horizon of the planning period or a uniform demand for the components. Anyway, between two program meetings, the system can not absorb any fluctuations in demand. Flexibility must then be served by buffer stocks.

15.6.5 Optimized Production Technology (OPT)

About 1980, Eliyahu Goldratt developed optimized production technology (OPT) as a Western answer to the Japanese introduction of JIT. The technique is based on a software that schedules by taking into account limited resources and any other constraints that would affect a firm's ability to adhere to a schedule (Chase and Aquilano, 1992). The scheduling logic is based on the separation of bottlenecks and non-bottlenecks and can be run on a computer in a fraction of the time that a MRP system uses. Since the whole production is balanced after the bottleneck resource, the concept is also referred to as Synchronous Production.

Figure 15.17 illustrates a linear flow of products with a capacity constrained resource. The bottleneck resource is a constraint for the whole production system, and the OPT theory is based on the requirement that only the bottleneck resource has to be scheduled separately to satisfy the marked demands. All non-bottleneck resources are then treated in a way to serve the bottleneck resource only. Production orders to non-bottlenecks in front of the bottleneck resource are initiated by a pull from the bottleneck. After the bottleneck, the operations are dispatched by the means of a pull through the remaining production chain. Buffer stocks are used in front of the bottleneck resource to secure that the bottleneck never goes out of

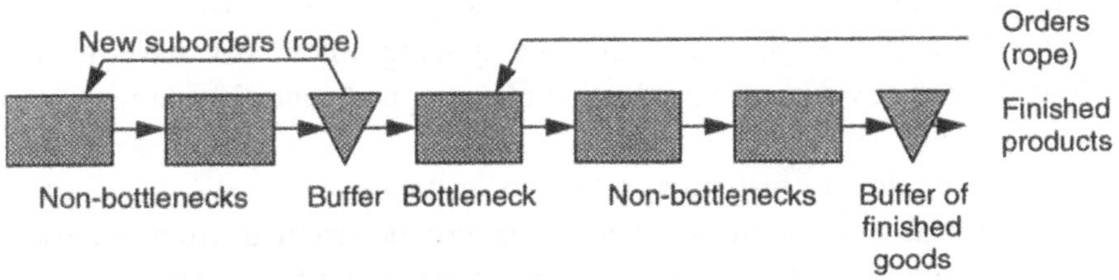

Figure 15.17 The OPT principle.

work. In the same way, there has to be a buffer stock of finished products to make up for deviations in the production and provide flexibility to meet changes in the customer demands.

The OPT philosophy is based on two sets of measurement that must be used together to measure the performance of a manufacturing firm. The financial measurements are:

- *Net profit*, which is an absolute measurement for the total return of the enterprise.
- *Return on investments*, which is a relative measure of the return in proportion to the investments.
- *Cash flow*, which measure the firm's ability to generate cash for daily survival.

At the operation level another set of measurements must be used as guidance:

- *Throughput*, which is the rate at which money is generated by the system through sales. It is important to recognize that only products that are actually sold, give a positive contribution to the profit.
- *Inventory*, which is all the money that the system has invested in purchasing things it intends to sell. This means that all raw materials, work in progress, and stocks of finished products are treated as inventory and do not generate profit until it is sold.
- *Operating expense*, which is all the money that the system spends to turn inventory into throughput.

The essence of the concept is that no value is added to the inventory before it is sold and transforms into income. This highlights the fact that working with unsold products only generates operating expenses and creates an incentive not to produce any components before a demand exists. To improve the productivity of an OPT firm, the goal is to increase throughput while simultaneously reducing inventory and operating expenses.

The OPT philosophy includes nine guiding rules that have to be followed when developing a production planning and control system:

1. Balance the flow of materials and not the capacity.
2. The utilization of a non-bottleneck is not determined from its own capacity potentials, but from other constraints within the system.

3. Utilization and activity for a resource are not the same. Optimal utilization of a non-bottleneck may include idle time.
4. An hour saved at the bottleneck saves an extra hour to the entire system.
5. An hour saved at a non-bottleneck is a mirage and only adds an extra hour to its idle time.
6. The bottlenecks determine both the throughput and the inventory of the system
7. The batch sizes for transport do not have to be equal to the batch size for manufacturing.
8. The batch sizes for manufacturing should be variable, both for the different resources and during the time.
9. Preference can only be stated when all considerations are taken into account simultaneously. The lead time is a result of the schedule.

OPT emphasizes utilization of the limited resources. This means that only components that are demanded will be produced and create a pull of the production orders through the value chain. It is a concept to synchronize the production to obtain a uniform flow of materials. A limitation is that a bottleneck must be identified to take advantage of the technique.

15.6.6 Load oriented production

The concept of load oriented production (LOP) focuses on the resources load situation that forms the basis for the dispatch of orders. The concept was developed by H. Kettner and is further developed by H.P. Wiendahl. The LOP technique has acquired a considerable acceptance in Germany in the late eighties.

The LOP system is based on the funnel model illustrated in Figure 15.18. The funnel describes the coherence between the arrival, departure, inventory, and throughput time for a resource. Arriving orders make an inventory of waiting orders that through the manufacturing process, pass the funnel opening. The opening of the funnel represents the capacity of the resource, and the throughput time can either be affected by controlling the capacity or the arrival of orders.

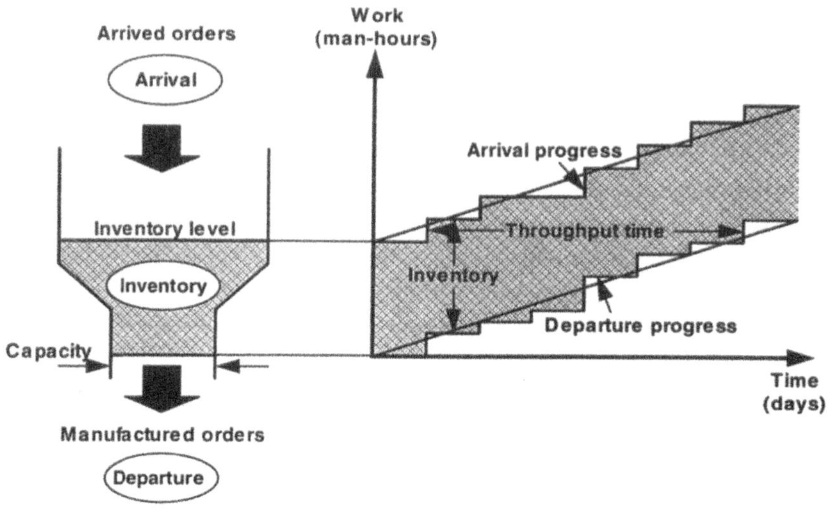

Figure 15.18 Funnel model and throughput diagram (Wiendahl, 1987).

The concept uses the coherence between throughput time, inventory, and capacity to control the throughput time. Since the capacity normally is given, the problem is only to manage the arrival of the orders in time with the departure orders so that uniform inventories and throughput times are obtained. The funnel is treated as a stable buffer stock. When a component has finished the manufacturing, a new component is ordered to satisfy the fixed inventory level.

15.6.7 Simulation

The latest development of computer power makes simulation a useful tool in production planning and control. When building a model of the production process, the computer can simulate the results of alternative strategic and operational decisions. Even if few enterprises utilize the possibility of simulation, a study (Bredrup, 1993) has shown that simulation of different production schedules gives a good overview and improves the results of production planning and control.

15.6.8 Developing an individual production management concept

It is impossible to propose a common production management concept that works for all industries. A company has always some unique characteristics and often exist in different market and competitive environment. Each company must choose an individual concept that fits their environment. For some enterprises, a standard system can be used directly or with small modifications to adjust to some specific need. Other companies can gain competitive advantage by development of tailor-made concepts.

Different market and competitive situations require the enterprises to emphasize dissimilar key parameters, and the described concepts support the key parameters in dissimilar ways. The MRP technique was developed on the basis of the push technique to schedule the production under the assumption that the lead times and demands are known in a long-term horizon.

Periodic Batch Control (PBC) is a simplified system that is based on fixed order cycle. Only the required orders for the next order cycle is manufactured in the current time period.

JIT realized with the *kanban* technique, is a practical concept using the pull technique. This approach requires uniform demand and a certain level of demand. If not, the required buffer stocks will grow too large.

OPT realize the advantages of basing scheduling on bottlenecks. If a fixed number of bottlenecks can be detected, this approach gives an efficient method of providing a schedule that makes the most of the bottlenecks.

LOP is based on a uniform load of the resources. The pull technique is realized by means of keeping a constant level of inventory in front of the resources.

In complex environments, the only usable method to choose a working schedule is by the means of simulation. Even if this approach is costly and time consuming, this method gives good results.

Several of the techniques can also be mixed. An enterprises that use a MRP system, can control several of their components that have uniform demand with a pull technique. This means that production chains that deliver components to MRP-based assemblies, can be managed with a *kanban* system.

The OPT model can also be used to separate a push and pull system. In front of the bottlenecks, the bottleneck can pull components by using *kanban*, and the components manufactured in the bottleneck can enter into the MRP system that manage the demand from the market.

16. PROJECT MANAGEMENT

16.1 PROJECT STAGING AND PHASING

Industrial production started based on projects. Each house constructed, each bridge erected, each automobile manufactured, etc., was tailored to the customer and performed as a one-of-a-kind task. Henry Ford with his assembly line and mass production technology changed this. Products and parts were made in batches or even continuously.

The project approach to production may therefore be regarded as the original one. It is still the dominating one in most construction businesses and in research and development. It is gradually also penetrating into the manufacturing industry. In many ways project-oriented work seems to be the future way of production. We shall therefore cover some of the most used techniques for planning and controlling projects. We will refer to this as project Management.

There exists many definitions of a project. Some prefer a vague definition such as 'something with an ultimate goal'. We shall use a more precise definition (Granli, 1986, Rolstadås, 1990):

> *'A project is a one-of-a-kind task with a clearly defined goal to be reached within a given time and cost frame'*

This definition directly highlights the three main control variables of a project:

- Scope of work
- Time
- Cost

A plan for these variables is referred to as the project control baseline.

Almost any individual in industry or business has in some context experienced project work. Roughly, projects can be classified into two categories:

- internal development projects;

- project production.

Internal development projects may cover any internal activity of an enterprise. Examples include development of a new product, implementation of new technology, construction of a new plant, study of productivity improvement, etc. Many of these projects are relatively small projects, and they may be executed by internal resources, external vendors or consultants or any mix of these two extremes.

Project production includes the situation where the company's end product to the customer is a project. An engineering company is typically an enterprise 'selling' projects. Another example is a research institute.

There are some important characteristics of a project which may help explain why projects will be the future mechanism of managing production:

- projects are goal-oriented and thus help to keep focus on what is relevant;
- projects are one of a kind and thus prevents the establishment of procedures, systems or habits that may be unnecessary;
- projects are by nature multidisciplinary and in this way promote communication across the organization of the enterprise.

In addition to this, we have to observe that there are a number of well-developed techniques and systems available on the market that can effectively help the planning and controlling of a project.

Any project may be split in three stages (Granli, *et al.*, 1986, Rolstadås, 1990):

- project identification;
- project definition;
- project execution.

The first stage is aimed at identifying possible prospects and doing an evaluation of these with the aim of finding a project that is feasible. For an internal improvement project this may include all brainstorming activities to find a prospect and the subsequent discussion and evaluation of these ideas to find the ones that are worth investing in. For an oil development project the identification stage would include exploration and reservoir evaluation.

The second stage would include conceptual engineering of the facilities or systems which are the goal of the project. In addition, this stage includes project planning. This means breaking the project down into manageable activities and providing a schedule and a cost estimate for each of these. We have then established what we just referred to as the project control baseline.

The third stage would be to execute the project, i.e. - to reach the project goal. This stage also includes monitoring and controlling the progress of the project as well as continuously revising the control baseline.

Each stage may be subdivided into a number of project phases. The actual number is dependent on the nature of the project and its size. For a small project the split in the three stages may usually be sufficient.

Each stage/phase represents a decision point in the project. The decision is to continue, to stop or to revise the goal of the project. Each stage/phase will therefore normally result in a report with a recommendation. This report must include a plan with activities, schedule and a cost estimate. Consequently there are several such plans during the project life time, each at a different accuracy level. This calls for a need to classify these plans (schedules and cost estimate) according to stage/phase, purpose and accuracy level.

Project management involves:

- project planning;
- project control.

Project planning will result in a project execution plant that as mentioned will form the baseline for project control. This project control baseline will contain three elements we have already indicated:

- work breakdown (MCWBS);
- cost estimate (MCE);
- schedule (MCS).

The abbreviation MC here refers to 'master control' which really means that it is the first original plan. Master control plans may be revised if the scope of work is changed. Otherwise plans are revised on a continuous basis all the time predicting final results. These plans are then referred to as 'current control' (CC) plans, i.e. CCWBS, CCE and CCS.

Figure 16.1 summarizes the concept of stages and phases.

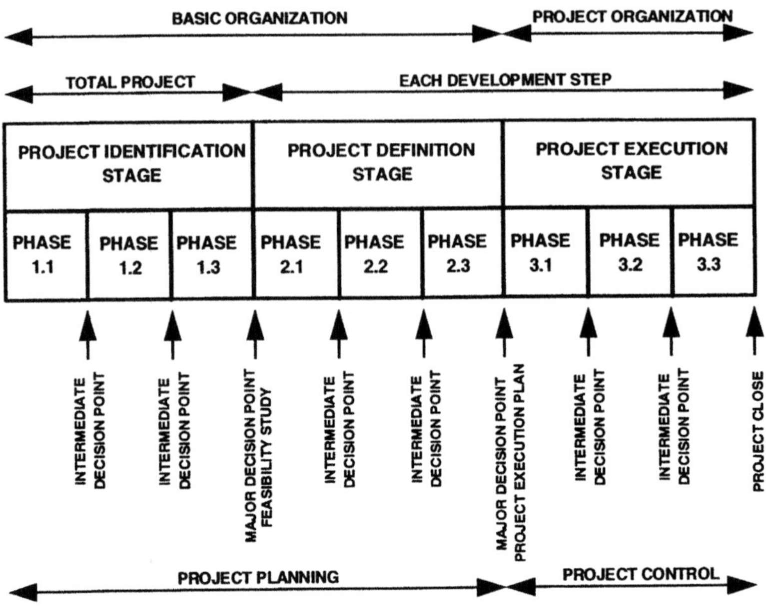

Figure 16.1 Project stages and phases.

In the succeeding we shall briefly describe the current state of the art for planning techniques and we shall indicate some of the main principles we can apply to control a project.

16.2 PROJECT IDENTIFICATION

16.2.1 Prospects identification

By a prospect we understand an idea or a possible project (Rolstadås 1990).

Productivity improvement is usually dependent on a flow of prospects in the enterprise. These prospects may emerge in all different parts of the organization, from the top management to the shop floor.

In many enterprises it is a major problem to identify the necessary amount of prospects. Stimulating measures to create more prospects is therefore of vital importance.

Most companies have boxes put all over where the employees can submit their suggestions for improvement. In many cases, the suggestions solicited in this way have proven insufficient. Other mechanisms must therefore be provided to compliment this. The Japanese '*kaizen*', often including quality circles, may be one approach. These quality circles meet on a regular basis to deal with current small and large problems as well as to discuss goals and means of further improvement.

A productive organization is much dependent on incentives to create prospects. Knowing the prospects, there is a need for methods to evaluate these and to take a decision on which projects to run.

Probably, the strongest incentive to obtain prospects or ideas, is the existence or awareness of a crisis. An accepted opinion in the organization that 'this company needs to change' will stimulate the creativity. However, changing by crisis is an insufficient technique. The need for a continuous change to reach agreed production and development goals must be encouraged.

There are numerous techniques available to develop prospects. One of these is scenario development. Through scenarios the future situation is predicted and the need to change from the current position becomes apparent. Another technique is brainstorming which is a creative session to obtain ideas. All possible ideas are recorded without any criticism or discussion during the sessions. Afterwards an evaluation is made.

In the succeeding paragraphs we shall assume the existence of a feasible number of prospects and we shall discuss how they can be evaluated and developed into projects.

16.2.2 Prospects evaluation

The easy way to evaluate a prospect is to look at financial indicators defining the project's profitability. This is also the most applied approach. It is of course a necessary condition, however, it may not be a sufficient condition. Other factors must as well be taken into account. The technical solution may be one such factor that should be taken into account. How technically feasible is the project? Should we select option A or B?

There are many such factors that need to be evaluated. They are all factors that rely on some subjective judgment in contrast to exact objective measures such as the financial indicators. We will refer to such evaluation criteria as intangibles. Intangibles can be literally anything from work environment through political image to technical feasibility.

Of course, evaluation of intangibles, is subjective. The usual procedure is to list a number of factors and to give scores to each factor. The factors are then weighed together to give an overall score or evaluation. Table 16.1 shows an example.

Table 16.1 Example of evaluation of intangible factors.

Factor	*Weight factor*	*Score*	*Weighted score*
Technical feasibility	3	6	18
Extension possibilities	3	2	6
Public opinion	5	3	15
Pollution	4	4	16
Flexibility	2	2	4
Total			59

The most used economic indicators are (Newendorp 1975):

- Pay back time;
- Return on investments (ROI);
- Net present value (NPV);
- Internal rate of return (IRR).

The pay back time is defined as the time elapsed from investment until the investment has been paid back. This is a very rough measure that should only be used in conjunction with some of the other indicators.

ROI is the ratio between the total proceeds from the project and the investment. It may be based on discounted or non-discounted values. This criterion is in frequent use for investment projects.

NPV is the discounted net value of the project, usually at the start-up point in time. NPV is calculated by considering the time-phased sequence of all incomes and all reimbursements (or expenses). The difference for each year is discounted to year '0' using an interest rate r. The interest rate, r, then represents the requirement for payback on the project. Projects with NPV > 0 are considered profitable whole projects with NPV < 0 are unprofitable. The interest rate is usually selected as the market rate or the rate obtained by alternative investments in the open market.

The most popular criterion is probably IRR. IRR is defined as the interest rate giving NPV = 0. IRR must be computed by an iterative algorithm. The benefit is that it is not necessary to stipulate an interest rate. The project having the highest IRR is considered the most profitable.

A sound project evaluation is seldom based on one of these criteria alone - usually several criteria will have to be considered.

To combine economic criteria with intangibles is possible by transforming the economic indicators to a score according to some defined scale. Then the scores from the economic indicators can be weighed together with the scores from the intangibles.

16.2.3 Decisions under uncertainty

Most project evaluation techniques are based on deterministic figures. However, this is rarely the case in reality. All estimates and figures are associated with some uncertainty. It is desirable to assess this uncertainty and take the corresponding risk into account for the project 'go' decision. There are several ways to take risk into account:

- Risk-equivalent values
- Expected values
- Sensitivity studies
- Stochastic planning

Risk-equivalent values compensate for an associated risk by giving a lower or higher figure than the deterministic one. This figure will then include a security against a loss due to the uncertainty. For example a forecasted income can be set lower to compensate for uncertainty. The risk-equivalent values can be computed by known statistical techniques.

A different approach is to use the expected value instead of a single deterministic value. However, this requires that the statistical distribution is known, at least empirically.

The third possible technique is sensitivity analysis. This assumes a deterministic calculation. After that each variable may be considered stochastic and the sensitivity of the performance measure is calculated by varying this parameter. The technique only allows to study one variable at a time. Simultaneously varying several parameters is only possible by using stochastic planning. All parameters are then assumed to follow a statistical distribution. By simulation the resulting distribution on all performance measures can be computed empirically.

There is no shortage of statistical tools to handle decisions under uncertainty. The problems in practice, however, are:

- How to assess uncertainty?
- How to communicate uncertainty?

For tasks carried out repeatedly, uncertainty can be estimated by obtaining a set of figures from past experience. However, often such figures do not exist. Then the most common technique is to use three estimates:

a - optimistic
m - most likely (realistic)
b - pessimistic

By assuming a β-distribution (which may be skewed), expected value and variance can be computed.

Even though uncertainty can be estimated, there is still the problem of communication in the organization. Few people in an enterprise are trained to think of data as stochastic variables. Many questions arise such as 'What does uncertainty mean?', 'How shall we use uncertainty?', and 'How can we assess and influence uncertainty?'. The answer of course, lies in better knowledge and training.

16.3 PROJECT DEFINITION

16.3.1 Work breakdown

One of the most difficult tasks in project planning is to break the project down into manageable units or activities. This breakdown is referred to as work breakdown structure (WBS). Few guidelines and almost no theory exist for this. This is the art of the engineer doing the planning. However, it does have major impact on success or failure of the project.

A WBS is a hierarchical breakdown of physical units (products/parts) and tasks (work activities and services) that are necessary to complete the project (Rolstadås, 1990). The number of levels will depend on the size of the project. We have shown an example in Figure 16.2. We have indicated four levels:

1. Project
2. Control center (CC)
3. Contract package (CP)
4. Work package (WP)

There may be several intermediate levels between each of these four main levels.

The levels may be associated with various aggregation levels of schedules and cost estimates. Also the top levels will usually correspond with the project organization, at least for a large project.

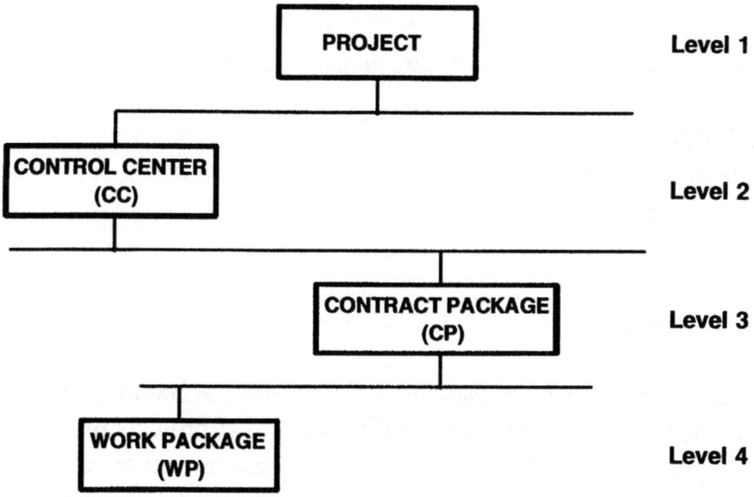

Figure 16.2 Work breakdown structure.

The control centers (CC) are overall responsibility areas. These may well correspond to organizational responsibilities in projects with several owners. This is usually the level of reporting from the operator to the partners. Control centers will often follow a breakdown of the final product in major components.

It is assumed that each control center involves a number of contracts (external or internal). This level is denoted the contract package level in Figure 16.2. Each contract may again include a number of work packages (WP). A WP is the smallest planning unit in the project, and represents the main planning level. For each WP the following data is planned:

- cost
- time
- resources

This is the crucial planning data. A common practice is to assemble the data on a form also including a description of the scope of work. Such a form is called a CTR sheet and includes all the data for that WP in the project control baseline. The collection of CTR sheets is referred to as a CTR catalogue and contains the complete project control baseline.

Within each WP resource is specified according to a cost breakdown - a code of account. For planning purposes it will also be necessary to specify activities in the work package.

As mentioned the design of the WBS is of crucial importance. By its design some important decisions are taken, such as:

- work flow;
- contract strategy;
- control and reporting level.

16.3.2 Scheduling

Scheduling of projects is usually different from scheduling of batch or continuous production. The main difference is that projects usually require that precedence relationships between activities are more important and therefore network type of techniques has to be applied. Of course, the most used scheduling technique is the bar chart or the Gantt chart. No diagram has ever conquered this in its ability to effectively communicate a schedule. Its weakness is of course the lacking ability to consider dependent activities. This aspect is well handled by networks. The output of a network, may however be communicated by a Gantt chart.

The classical network approach comprises two techniques:

- CPM - Critical Path Method;
- PERT - Program Evaluation and Review Technique.

CPM works with deterministic estimates of activity duration whilst PERT is capable of handling stochastic activity duration.

Networks are represented by graphs and two possible approaches exist:

- AOA - Activity on Arc;
- AON - Activity on Node.

Figure 16.3 shows examples.

The purpose of a network planning is:

- to determine the project duration;
- to determine which activities that influence project duration.

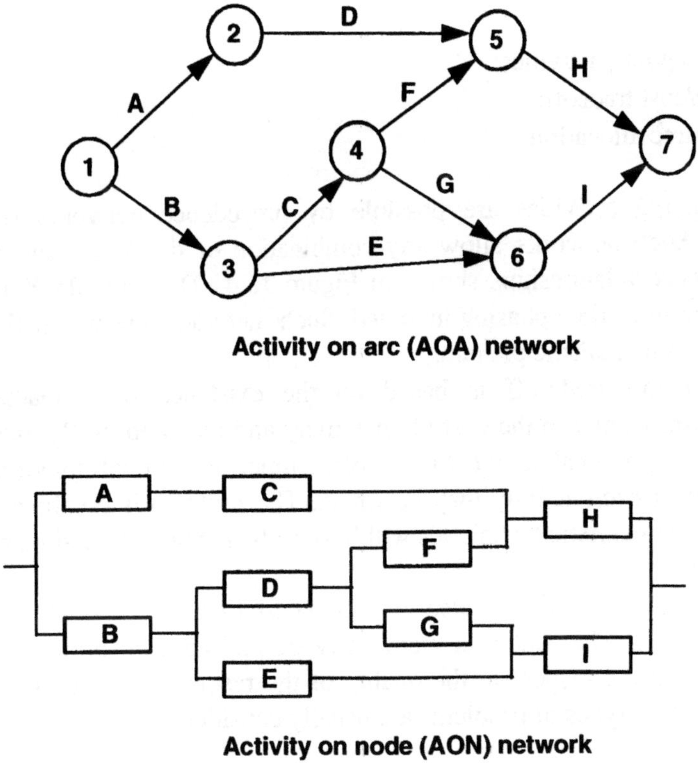

Figure 16.3 Examples of network representation.

Calculation is done in two steps (Lockyer, 1984): a forward and a backward one adding and subtracting the duration respectively. In this way the following data for each activity is fixed:

- ES - Early Start
- EF - Early Finish
- LS - Late Start
- LF - Late Finish

In addition 'float' is determined. Float is the flexibility we have to move an activity in time without affecting the project duration. Activities with zero float are critical. A chain of critical activities from start to finish represents a critical path.

The classical approach to network scheduling has later been extended to cover other types of problems. The most important extensions are:

- overlapping activities;
- time/cost tradeoff;
- resource allocation.

Overlapping activities are possible by precedence networks (Lockyer, 1984). Such networks allow any combination of the four basic types of precedence relationships shown in Figure 16.4. The variable X indicates the overlap or time phasing involved. Such networks are useful since they allow a more realistic planning.

Time/cost trade-off is based on the existence of a mathematical relationship between the cost of an activity and its duration. By associating a cost proportional to the total project time, an optimal solution to the duration of each activity may be found. The problem is combinatorial by nature, and heuristic techniques will have to be applied to obtain a practical solution.

Resource allocation is also a combinatorial problem and likewise calls for heuristics. The quality of many software packages for project planning, differs with the respect to the quality of the resource allocation algorithm applied. Two types of problems are usually considered:

- resource smoothing;
- resource allocation.

The two main variables in such a resource problem are the level of

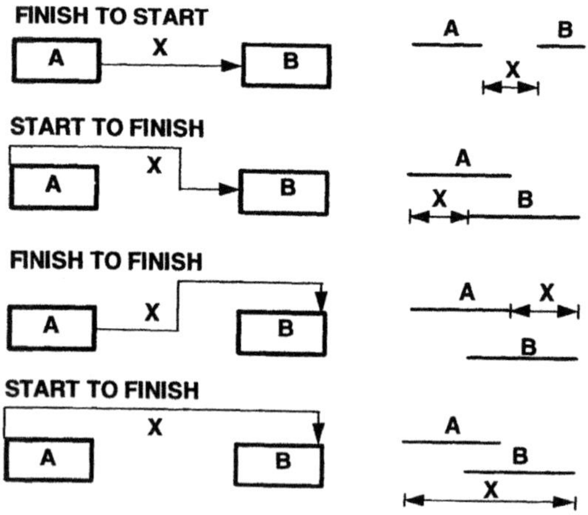

Figure 16.4 Precedence relationship.

available resources and the project time. Only one of these parameters may be kept fixed.

In the resource smoothing problem the project duration is fixed and the resource level is kept as low as possible. In the resource allocation problem, the resource level is given and the project duration is minimized.

In both cases the flexibility in planning is first of all in utilizing the available float in the network.

16.3.3 Cost estimating

In chapter 16.1 we looked at the various stages and phases of a project and we stated that each phase represents a decision point. The decision is taken based on schedule and cost information amongst others. Actually, during a project life time, there are several cost estimates defined, each different in detailing level and accuracy. We can use this information to classify the estimates. If we consider a small project with subdivision only on stages, we will consider the following two categories (Clark and Lorenzoni, 1985):

- A - Screening estimate
- B - Control estimate (MCE)

The rationale for classification of estimates is that we know in each case the technical documentation that the estimate is based on as well as the associated uncertainty.

In the example above, the B estimate is the estimate included in the project control baseline. We refer to this as the master control estimate (MCE). This is also the estimate we referred to at the work package level in section 16.3.1.

During project execution this estimate is regularly updated. We refer to the latest version as the current control estimate (CCE).

We usually distinguish between two types of estimating techniques (Rolstadås, 1990):

- synthetic
- analytic

The synthetic techniques are applied in early phases to provide rough estimates. They are made from vague engineering definitions and do not require a detailed project breakdown. The accuracy level may be in the range of ± 20%.

There are two types of synthetic technique (Granli, *et al.*, 1986, Rolstadås, 1990):

- relational estimating;
- factor estimating.

The relational estimating is based on a previous similar project with a known cost. The assumption is that the new project will have the same cost. However, corrections are made to compensate for:

- difference in capacity;
- different location;
- different time.

Factor estimating is based on the assumption that the relative share of various cost categories are invariant. Estimating one category then gives the total cost split on the categories if the shares are known. Analytical estimating is based on a detailed project breakdown. For each activity or unit, costs are estimated, usually based on bill of quantities. These bills of quantities are taken from engineering drawings. Unit rates and prices are retrieved from an estimating database.

The bill of quantities that the cost estimate is based on is associated with some uncertainty. The further the engineering has progressed, the more accurate the figures are. However, at any stage, we want to estimate the final cost. To compensate for this uncertainty we introduce a cost contingency (Rolstadås, 1988). This shall cover unexpected things that we know will occur, but cannot identify at the time we are making the

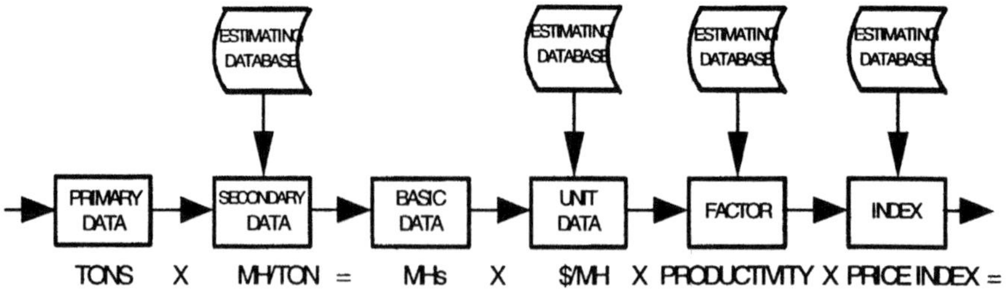

Figure 16.5 Example of estimating chain for labor cost.

estimate. Contingency is therefore a cost item that will be gradually spent during the execution of the project.

The calculation of a project contingency is a matter of great concern. A theory has been developed based on the assumption that the estimate provided by an estimator is the most likely value. For control purposes, however, a fair estimate should have equal chances of overrun and underrun. In statistical terms this means the median. In this context the median can be approximated by the expected value which is more convenient to compute. Assuming that the estimated costs follow a skewed distribution as depicted in Figure 16.6, the contingency can be defined as the amount that must be added to the base estimate to obtain a 50/50 probability. The base estimate is then defined as the most likely value provided by the estimator based on the existing drawings.

In this terminology, half the projects should exceed the budget and half underrun it. If a reserve to cover an overrun is necessary, this should be added in addition. Whereas the contingency is expected to be spent, this management reserve is likely not to be spent. A budget could then be composed in the following way:

  Base estimate
\+ Contingency
= Control estimate
\+ Management reserve
= Budget

There is much confusion concerning contingencies and risk in cost estimates.

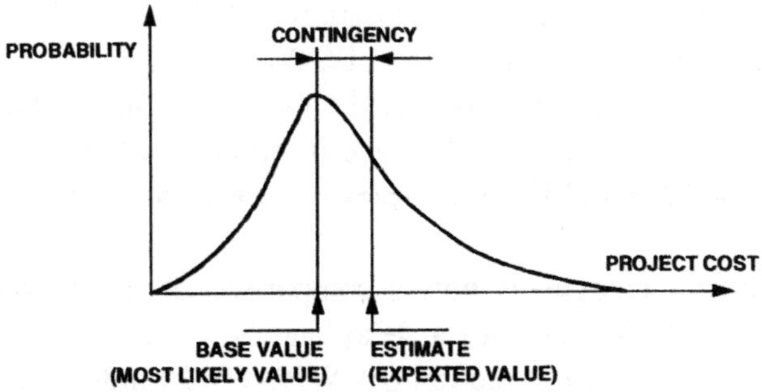

Figure 16.6 Base estimate and contingency.

16.4 PROJECT CONTROL

Project control will guarantee that plans are followed with respect to scope of work, time and cost. We will here limit ourselves to discuss some of the main principles applied in project control. These fall into six categories (Rolstadås, 1990):

- Earned value;
- Productivity measurement;
- Resource availability;
- Physical progress;
- Tendency development;
- Forecasting.

The principle of earned value involves a measurement of what has physically been completed. We compare this to the planned value. Thus we obtain a real measure in volume of work of how far we have progressed compared to the plan. We have to options, either to measure man hours (earned hours) or costs (earned value).

Productivity in this respect is defined as the ratio between planned and actual resource consumption. Productivity is therefore a measure of how close to the plan our real progress is. We can measure both time or cost. We can report productivity performance by indexes or by a progress diagram.

Figure 16.7 shows a progress diagram based on costs. It shows the estimated costs at work scheduled and earned value (estimated costs at work performed). In addition the actual cost of work performed is shown. The productivity is the ratio between the curves ECWS and ACWP. At the time t, the total cost deviation is the difference between ACWP and ECWP. This is the sum of a budgeting deviation (ACWP – ECWS) and delay (measured in work volume) (ECWS – ECWP).

A full project control must be based on the three curves shown in Figure 16.7. However, this is seldom the case in practice. Many companies only consider the ACWP and ECWS curves, i.e. they focus only on the budget deviation.

We have now recognized how a comparison of earned value (ECWP) and the plan (ECWS) indicates a delay and how the ratio between actual and planned resource consumption (ECWP/ACWP) indicates productivity. A poor productivity may be due to the fact that the work requires more

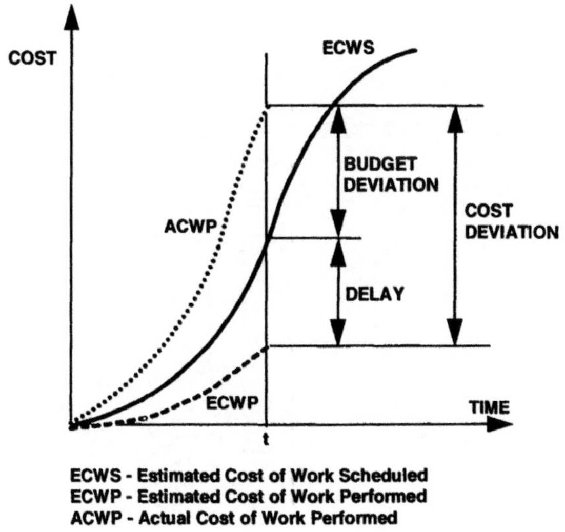

Figure 16.7 Example of progress diagram.

resources then what was anticipated. The reason for a deviation in volume may be due to insufficient availability of resources. It is therefore important to monitor the resource availability. This is usually done by a histogram showing actual versus planned cumulative manning.

The progress diagram we have used so far gives no information on schedule progress. The best way to do this is by comparing 'milestones'. We call this monitoring the physical progress.

Most companies look at cumulative or aggregate figures when controlling a project. However, the first derivative , i.e. the tendency, gives better information. It opens a possibility for the management to react on a deviation at a much earlier stage.

These tendencies may also be used to forecast where the project will end with respect to cost and schedule. This type of information is usually very valuable since it increases the awareness of reaching the project goal.

17. PURCHASING

In common situations, purchasing can have different meanings, depending on how, and by whom, the word is used. For example, purchasing can mean a specific department within a firm, a transaction done by a firm to acquire a certain goods, or a type of action related to acquisitions of goods in general. To be able to say anything about purchasing in general, our first task is thus to define respectively purchasing, strategic purchasing, the purchasing function and the purchasing department:

> *By purchasing, we will think of all those decisions and activities in a firm related to the task of acquiring goods, services and information congruent with the firm's goals.*
>
> *By strategic purchasing, we will think of those purchasing activities related to the long-term development and maintenance of suppliers and supply sources in order to achieve competitive advantages.*
>
> *The purchasing function is defined as the functional area inside a firm that carries out purchasing activities as defined above. In this function is included all persons that partake in at least some purchasing activities (Kroken and Pedersen, 1992).*
>
> *The purchasing department is the organizational unit or units inside a firm that are responsible for some or all of the purchasing activities.*

There are two important things about these definitions. The first thing of importance is that our definition of purchasing includes the informational aspect. Late works on purchasing (Farmer, 1985; Scheuing, 1989; Gadde and Håkansson, 1993) show that suppliers are a very valuable source of useful information. In particular, they are important when it comes to information about technological development, products and product development and the supplier market in general.

Another important aspect of these definitions, is the difference between the purchasing function and the purchasing department. As is evident from our definitions, the purchasing function comprises more than the purchasing department. The reason for introducing this difference is that different firms apply a range of organizational solutions to the problem of organizing purchasing activities. Thus, any firm must perform the purchasing function, but not all firms have a separate purchasing department.

17.1 WHY EMPHASIZE PURCHASING?

Lately, purchasing has been focused on a lot, both in the private and public sectors. Ericsson and Persson (1986) have in an article identified the current industrial situation as a material flow-oriented stage where the purchasing function is given a central role. Another important aspect is that the strategic purchasing functions are emphasized. If the authors of this article are right, many firms will find themselves in a position where they need to upgrade their purchasing function.

Scheuing shares the view of Ericsson and Persson. He claims that the position of the purchasing function has developed from an internal cost center and a necessary evil, and into a profit center that in itself can be a strategic, competitive advantage.

The shift in the perception of the purchasing function identified by these authors, can be seen as a result of several factors that together have served to change the conditions for purchasing activities. One of the strongest factors may be changes in production technology. The increase in automation and the corresponding increase in employee productivity has caused a change in the total cost structure of the typical firm. In earlier periods, wages and other costs of hiring labor were the dominating costs of doing business, and industries were labor-intensive (Kroken and Pedersen, 1992). Today, costs of materials and materials handling represent 50-80% of the total costs of a typical industry enterprise.

Brenden (1990) elaborates on this economical argument. He claims that a survey among 200 Norwegian industrial firms, shows that almost two out of three respondents had purchasing costs that constituted at least 63% of their sales. Research done by Håkansson (1989) shows similar results for Swedish companies. This research also concludes with the interesting observation that for an individual firm, a very small number of suppliers often constitute 50-80% of the total purchasing costs.

The appertaining economic importance of the materials costs should lead to increased interest in this area. One can easily see that a 5% reduction in purchasing costs leads to a substantial improvement on the balance sheet. In fact, a company where materials costs are 50% of total sales, would need to improve it's sales record by 25% to *increase the equate balance* to a 5% reduction in purchasing costs.

The resulting argument is that purchasing is developing as a functional area within a business enterprise, and is rapidly becoming more important as a strategic tool. It needs to be emphasized that a strong focus on strategic and long-term aspects of purchasing, will increase the company's ability to compete and lead to short-term economic benefits as well (Cammish and Keough, 1991).

17.2 IMPORTANT DIMENSIONS IN PURCHASING

Traditionally, the purchasing function's task has been to take care of ordering of raw materials. To this simple task, there were quickly added the task of checking the received goods including quality controls of incoming goods and all paperwork regarding contracts with suppliers, shipment orders, custom bills and other necessary documents. Later, other tasks have been continually added to the responsibility of the purchasing function.

To make some systematic thoughts on the task structure of a modern purchasing function, we must therefore choose some dimensions to make certain categories. We have chosen two dimensions; the degree of strategic elements in the task, and whether the task is directed inwards to the rest of the company or outwards mainly toward suppliers. This forms a 2x2 matrix that we can use to categorize the different tasks that the purchasing function should be responsible for (Pedersen and Torvatn, 1993), see Figure 17.1.

	OPERATIVE	STRATEGIC
INTERNAL	Purchasing routines	Purchasing organization
EXTERNAL	Supplier routines	Supplier strategy

Figure 17.1 Purchasing tasks.

We define four fields that the purchasing activities can be categorized into:

- Operative, internal: purchasing routines
- Operative, external: supplier routines
- Strategic, internal: purchasing organization
- Strategic, external: supplier strategy

17.2.1 Purchasing routines

These tasks are the day-to-day operations of the purchasing function. Usually, they are handled by the purchasing department, but they can also be diffused among line units. These tasks can generally be divided into 12 areas (Kristiansen, 1991):

1) **Awareness of needs**. It is important that the company can discuss whether there really is a need to buy something. This area applies in particular to purchasing decisions regarding equipment and other types of goods not directly used in the end product.
2) **Soliciting bids**. In this phase, it is necessary to determine exactly what properties the needed goods/service should have. It is generally a good idea to determine criteria for choosing between bids in this phase.
3) **Evaluating bids**. The exact nature of the work in this area would depend on the number of bids received, but generally involves deciding whether each individual bid can fulfill the properties determined, and then evaluating the bids using the criteria determined in phase 2.
4) **Choosing a supplier**. Phase 3 has laid down the groundwork for this decision by evaluating the bid. Now, it is necessary to decide which supplier to negotiate with, and to decide whether we need alternatives. In many purchase situations, phase 2 and 3 are skipped because bidding is not necessary or not wanted. Long-term strategic choices and/or buying in a near-monopoly market are two examples where bidding is not a realistic choice.
5) **Negotiations and 6) Creating and signing contract(s)**. Whether bids have been solicited or not, it is often necessary to discuss and possibly negotiate on terms of the delivery. Price and what should be included in the delivery is also important, as well as possible provisions for developmental work, future obligations and conflict resolutions. A

contract will almost always be prepared, but the content of this contract can be very detailed, or very broad, depending on the nature of the purchase and the nature of the relationship between company and supplier.

7) **Requisitions and 8) Ordering**. In some cases, especially where repeated purchases are used, the purchasing department needs to handle requisition orders from other parts of the firm. The goods or the service may then be ordered. It is obviously of vital importance that the purchasing department have a thorough knowledge of all contracts and other deals with suppliers at any time. Thus, the ordering routines are among the most important operating routines in the purchasing function.
9) **Following up**. In this phase, the purchasing function is responsible for following up any deliveries not yet made. Deliveries deviating from the contract or deal in any way, should be followed up immediately.
10) **Control of deliveries**. Deliveries are often controlled upon arriving in the company. Such controls can be just a check on type and numbers, or it could be a serious quality control. In the latter case, it is usually handled by the quality department, and not by the purchasing department. Again, any deviations from contract should be followed up immediately.
11) **Controlling the receipt and 12) Paying the goods/service**. The freight documents and other receipts should be controlled by the person ordering the goods. The receipts can then be sent to the accounting function for further treatment.

17.2.2 Supplier routines

This is the administrative work related to the suppliers. We can sub-divide this work into three areas:

- developing performance criteria for suppliers;
- screening of the supplier markets;
- supplier evaluation.

Again, this work will normally be done by the purchasing department, but it could also be handled through other departments.

Developing performance criteria for suppliers

In this work, the task is usually to enable the purchasers to look at certain attributes in order to discriminate between suppliers that offer the same product or service. The pertinent question is: What do we want from our suppliers? Several authors (Cravens and Hoffman, 1977; Blom and Hovda, 1985; Farmer, 1985; Spekman, 1988; Brenden, 1990; Kristiansen, 1991) have tried to answer this question, and we will present a list of possible performance criteria that these authors have chosen as important. Criteria one to five are related to a specific exchange, while criteria six to nine are related to the supplier.

1) **Service**. Although the word service itself has been defined in several different ways by these authors, the common use of the word is how quickly can the supplier react to problems with the goods/service they have delivered. Good service means a quick response by the supplier, and that the supplier automatically accepts such problems as their responsibly.
2) **Quality**. Most important is that the delivery of goods is according to the standards set in the agreement.
3) **Timeliness of deliveries**. This is generally regarded as important, and is crucial for firms running on tight schedules. Just-In-Time production systems build on the principle of deliveries on time, and cannot afford to stop production waiting for deliveries. Other companies may or may not have some slack depending on the importance of the goods delivered, but it is almost always the case that a late delivery is a threat to the buyers' ability to arrange delivery on time.
4) **Price**. Corresponding to the authors mentioned, price is a very important aspect, although not the single most important. In most cases, price is a question of the right price, and not necessarily the lowest price. It deserves mentioning that the price itself is usually only the 'tip of the iceberg'. Administrative costs, transportation costs and other transaction costs are also factors that should be taken into account.
5) **Volume/quantity**. Again, it is a question of the right quantity. The importance of the right quantity is closely linked to timeliness of deliveries. An example is JIT production systems, which are very vulnerable to deviations in quantity.
6) **Technical competence**. Is it possible for the supplier to help the buyer with technical development? Is such help needed by the buyer? How good is the technical competence of the supplier? Answers to questions like these are often important criteria in choosing suppliers. In

particular, such aspects are important when buying machines and other equipment not directly related to the core competence of the firm.

7) **Economical aspects**. Here, it is important to evaluate the economic and financial position of the supplier. This should be done in order to make a guess on whether the supplier has the economical strength needed to be a valuable exchange partner.
8) **Geographical and cultural distance**. Handling suppliers in other parts of the country and in other countries may offer new challenges to the buyer. Administration costs, transportation costs and coordination costs tend to increase, and language and cultural differences tend to reduce the effectiveness of cooperation and technological exchange.
9) **Future plans**. The possibility of a future relationship with the supplier is important, and most authors emphasize that such relationships should be developed gradually. Still, it is reasonable to look for such possibilities even as you choose suppliers.

These criteria should form a solid basis of information for determining supplier strategies and can also be used on an operative level when the firm chooses between different suppliers in a bidding round. For practical use, it is important to weight the criteria, so that the information can be evaluated on a holistic level. All performance criteria are not equally important, and we should think through the consequences of failing to some extent in one or maybe even several of these criteria. Can we for example exchange volume for better service? The answer would vary widely depending on the company, the goods exchanged and the nature of the exchange situation.

Screening the supplier markets

Another important function in this field is the screening of the supplier markets. The company should stay informed on the development in their most important supplier markets. This means both general trends that could affect the way business is done in this market, and the development of specific firms and companies in this market. Such data needs to be collected systematically, and then stored and extracted into useful information to serve as a basis for decisions.

Scheuing (1989) lists a number of sources for such data:

- published sources;
    - ✓ general industrial registers or directories;
    - ✓ industry sources;
    - ✓ yellow pages;

    - ✓ manufacturer's catalogs and sales literature;
- internal sources;
    - ✓ supplier files;
    - ✓ technical personnel;
- personal contacts;
    - ✓ sales personnel;
    - ✓ professional colleagues;
    - ✓ trade exhibits;
- international sources;
    - ✓ department of commerce;
    - ✓ foreign embassies and consulates;
    - ✓ international chambers of commerce.

Supplier evaluation

Many authors recommend that a company performs supplier evaluations (Farmer, 1985; Scheuing, 1989). We do agree, but want to add that it is very important to think through the reasons for evaluating, and to determine how the evaluation is going to be followed up.

Relating the evaluation to the purpose

Evaluations can be done mainly for two purposes:

- in order to judge whether to keep or switch suppliers;
- in order to form a basis for discussions between supplier and buyer upon the performance of the supplier.

The form of evaluation chosen should reflect the main purpose.

In the first case, passing judgment is the important purpose, and the evaluation should be very formal, criteria-based and done in such a way that the supplier feels that it is a hard, but fair evaluation. Preferably, the suppliers should know on which dimensions they are being evaluated, so that they can try to improve on those dimensions.

In the second case, the main task of the evaluation is to provide information as a basis for the discussion. Thus, the evaluation should be more open in it's way of questioning, and function more as a checklist on important areas.

Methods for supplier evaluation

When choosing how to evaluate, there are several important methodical questions that should be true. In addition, to the purpose of the evaluation, there are three other crucial decisions to make:

- Which questions should we ask?
- How will we collect answers to these questions?
- How will we treat the answers?

When we talk about what questions, the relevant decisions are which areas we want to know something about. In general, we want a lot of information on the supplier's production capacity, technological competence and production economy. The problem is that the supplier is often interested in protecting this information, as it is the basis for their own value-added work in the production chain. Other information that might be useful is information on the supplier's general economic conditions, their staff, their connections to other firms and their future strategies. In all these areas, information can be more or less detailed.

It is obvious then, that a buyer cannot obtain full information on all possible suppliers. Therefore, the number and type of questions asked should reflect the importance of the goods wanted. If the goods purchased is a crucial input to the buyer's production, this warrants a very thorough analysis of the supplier. Adversely, if the goods are standardized and unimportant, it is probably a good idea to drop evaluations totally.

Collections of such answers are also a problem. One way is to make standardized forms and send them to suppliers, so that they can fill in their answers. This method is cheap and enables the buyer to reach a lot of suppliers. Answers are generally on a standardized form, and can easily be treated statistically through index construction (Cravens and Hoffman, 1977). The problem is that answers given in such forms are often inaccurate. Questions have been misunderstood, and/or the questions have not been answered by the right people (e.g. the supplier's sales/marketing department answering questions related to production and economy). Also, the supplier tends to arrange for answers that are better than they should be. In general, such methods are recommended for suppliers of secondary importance.

To cover such problems, it is possible to send representatives of the buyer to the supplier to help with the evaluation. Such representatives can often eliminate misunderstandings and make sure that the right people fill in the answers, but the costs of gaining information will obviously increase. Also, the value of the information gained is still in doubt since many of the criteria chosen are hard to quantify. While economic factors exist in a quantified state, they therefore are easy to access. There exists no such quantification of technological expertise or future strategies. These factors will always be a matter of judgment. Still, for the most important suppliers, sending representatives to make their own judgments and to discuss with the supplier, usually is the best option.

Finally, the question on how to treat the answers, needs some decisions. Statistical treatment is often used, but this requires very precise questions and very precise answers. Non-quantified information is hard to access in this way, and the buyer is forced to use surrogates for such information. One way to do this is to use scales where the respondent should indicate agreement with a certain declaration made in the form. The advantage is that statistical treatment is fairly easy to do, and can present the decision maker with relevant data in a way that is easy to use.

There are a lot of things to be aware of when doing external, operative work within the purchasing function. However, the most important thing to keep in mind is that you should not treat all your suppliers with equal detail. Key suppliers in the most important areas should be treated much more thoroughly than other suppliers.

17.2.3 The purchasing organization

As with most organization theory, there is no single way to organize the purchasing function. Rather, decisions upon how to organize the purchasing function should be looked upon as strategic decisions.

The purchasing department

It should be stressed here that no firm can avoid doing purchasing activities. If a firm does not have a purchasing department, it could mean that they have decided to split the activities between other groups and persons in the firm. More often though, the case will be that they have distributed some of the operative tasks, but that the strategic activities are all but non-existent. The first decision that needs to be made is the role of the purchasing department. What activities should it be designed to handle, and what activities should be left to other parts of the organization?

The main reason for task specialization in a firm is that a task requires special knowledge and/or that there is a similarity in work procedures, allowing collection of these procedures into one function. Purchasing has some problems establishing itself as a separate field of knowledge. Often it is viewed as a part of materials administration, of production or even as a part of the storage and handling function. Similarities in work procedures are there, but they are most obvious on the operative level. Thus, the arguments for task specialization in the purchasing function have traditionally appeared weak, and have thus often been ignored.

This has commonly lead to a very weak purchasing department, where only the operational tasks are included. Activities that we have included among strategic purchasing activities are in general handled by other managers. Also, the purchasing manager has been on the middle or lower management levels, thus effectively blocking him or her from partaking in the strategic work (Cammish and Keough, 1991).

We recommend that this should be changed. Due to the development of modern purchasing methods, the purchasing department should be strengthened by including strategic activities in it's area of responsibility. Obviously, the purchasing manager would then need to operate within the strategic top management group, at least on an occasional basis. This view is supported by authors such as Farmer (1985), Spekman (1988) and Scheuing (1989), who also argue strongly for a strengthening of the firm's purchasing department.

Another discussion is the placement of the purchasing department within the firm's organizational structure. Today's situation is that the purchasing department is usually on the 3rd or 4th level of the organizational hierarchy, usually as a subordinate department of either production or accountants' department, or both. Keeping in line with the argument above, we believe that the purchasing department should be in the 2nd level in order to be able to perform it's strategic activities unless there are very good reasons for not doing so. Integrating the procurement with other logistics activities at 2nd level to improve the material flow, is also desirable approach.

Blom and Hovda (1985) mentions several such factors that should be taken into account when making decisions on where to place the purchasing department:

- materials costs in absolute measures;
- materials costs in relation to total costs;
- the type of materials purchased;
- the strategic importance of purchasing.

Bondlid (1971) adds the following factors:

- the size of the firm;
- the type of industry the firm is involved in;
- the geographical placement of the firm's component parts, it's suppliers and customers.

Centralization or decentralization of the purchasing function

The purchasing department should generally be charged with performing strategic purchasing activities and elevated to a position where it's strategic activity was of significance. There exists another possibility that has been hotly debated in the literature, and that is whether it is better to decentralize the purchasing activities to the line managers. Bondlid (1971), Baily and Farmer (1990) and Kristiansen (1991) use the following arguments for decentralization of the purchasing function:

- The criteria for which goods to purchase should be set by the person who needs the goods. Essential information may get lost if the criteria are to be developed by a central department
- Decentralized purchasing reduces paperwork and communications lines, thus cutting time requirements
- Indirect costs may decrease depending on the geographical locations of the buying and selling firms
- Decentralized purchasing increases flexibility, because several limitations caused by centralization and coordination are removed

The same authors use the following arguments against decentralization:

- Coordination of purchases of the same type should bring better prices (and/or other benefits)
- Standardization of supplies and reduction of number of suppliers can only be achieved through centralization
- Indirect costs regarding ordering and transportation can be reduced through centralization
- Centralization increases the advantage of task specialization, and allows specialists to work on purchasing problems
- Control and planning routines are stronger when the function is centralized

Thus, there are good arguments both for and against decentralization. Each firm should decide the model that they want, but it is important to make such a decision explicitly and in regard to the specific situation and goals of the firm, rather than letting a mix of centralized and decentralized systems develop naturally.

17.3 SUPPLIER STRATEGY

Supplier strategies can be defined as:

> *Consciously employed actions to relate a firm's activities toward those of it's suppliers.*

17.3.1 Supplier development

Supplier development is a program or range of programs, initiated and supported by the buyer, which aims to develop some aspect of the operations of the supplier. This aspect can be financial, economical, organizational, technological or market-oriented, or it can be any combination of such. The aim for the buyer is to develop the supplier's ability to deliver first-rate quality for a good price.

Supplier development programs vary in size and cost. Very simple programs are usually concerned with the introduction of specific systems or standards into the supplier's organization. Examples of such programs are the introduction of new production machinery or certification for a new quality standard. Programs like these are limited in time, and the buyer's involvement is very specific to the task at hand. In some cases, the program can be linked up to similar programs within the buying organization, but it need not be so.

Due to the sheer costs and the drain on resources that such programs represent, they are usually enacted only toward key suppliers. What is more, they often have a long-term perspective, because of the need to recover these expenses. Thus, supplier development implies that the buyer has a strategic view of the situation. It becomes a question not only of what to do, but when to do it, and what suppliers to choose for such collaboration.

Since supplier development programs are strategic in nature, they are often performed by the top management in the company (if they are at all initiated). The purchasing department is often not involved in such programs, and if it is involved, it usually plays a very minor role. We do not agree to such a stance. More often than not, the purchasing function has important information about the suppliers. This information is better utilized if the purchasing professionals are included in the carrying out of such programs. To enable purchasing professionals to do this, they should be included in the strategic decision-making body of the firm.

17.3.2 Supplier/buyer cooperation: Strategic partners

The cooperative effort between supplier and buyer can be much more complicated than supplier development programs. Supplier/buyer cooperation is not unusual, and such relationships usually involve exchange of technological or market-oriented ideas, and often include adaptations of the production process. The result is savings in materials flow or lead times in production. One example is the introduction of new products, where buyer and supplier work closely together in order to achieve high-quality/low cost production of this product. Integrated production systems such as Just-In-Time systems, can be viewed as very complex forms of supplier/buyer cooperation, where the cooperation is not only between a single buyer and a single supplier, but involves a buyer and several suppliers, and necessitates coordination between the different suppliers as well since time is so essential.

Again, these programs are strategic in nature, and require a lot of investments to work properly. A necessary prerequisite for the success of such programs seems to be single-supplier strategies. This is when there is only one important supplier of a particular type of goods. Because of the long-term view and the high investments, it becomes too expensive for a firm to have dual sourcing in such situations. In today's companies, conventional wisdom has traditionally been that the firm should have several suppliers for each type of goods, and doubly so for key input to the production. The wisdom of this is now being questioned, and several firms have already changed to longer-term strategic partnerships for important sub-systems, components and raw materials.

17.3.3 Organizing supplier systems

Taking the supplier/buyer programs one step further, we can discuss whether such programs should also be initiated toward the supplier's supplier. In some cases it might even be beneficial for a company to organize and help develope a whole supplier system. We can see this happening today in heavy assembly industry, such as the car industry and the aircraft industry. Toyota, GM and Boeing are but three assembly firms who have helped in organizing and developing not only their closest suppliers, but whole layers of suppliers. In such cases, the closest suppliers (often termed 1st tier or 1st level suppliers) are themselves assembling components from subsuppliers and function as system suppliers in relation to the main assembler. As an example, Toyota has 94 such 1st tier-suppliers.

The main firm's chief function in the production system then becomes that of assembler of the finished systems into a product. The main firm also tends to handle distribution and marketing of the finished product. Product development is often handled together with the 1st tier suppliers and sometimes even with 2nd tier suppliers, but with input from firms much farther downstream in the system as well.

The organizations of multitiered supplier systems are extremely expensive, and only an option for the very strong business conglomerates. In Norway, only the assembly of oil rigs is really big enough for such supplier strategies. Yet, downscaled versions of the idea will become more commonplace. This might include system deliveries for key systems, where several tiers of suppliers might be involved.

17.4 COSTS, PRODUCTIVITY AND EFFECTIVENESS

17.4.1 Purchasing as a cost center

Most firms would say that purchasing is a cost center. This view of the purchasing function is partly correct, in so far as we choose to look at the operational parts of the purchasing function. Choosing this viewpoint, we can discuss three sources of costs in the purchasing function.

Direct costs linked to the specific transaction

The first source of costs is the price of the goods/service. Every goods/service acquired externally carries a certain price. In our country this price is usually paid in money.

Indirect costs linked to the specific transaction

The second source of costs is those indirect costs that are linked to a specific transaction. Obvious examples of such costs are transportation costs, customs costs and exchange rates and quality controls of incoming goods.

Recent literature, particularly the transaction cost theory (Williamson, 1985) and theory on quality management (Oakland, 1991), have pointed to other, more hidden costs as well. One such cost is the money lost due to lack of quality in the incoming goods. Lack of quality may lead to a lot of work invested in goods that is later thrown away. In extreme cases, such goods may even go all the way through the firm and end up at the customer to cause more trouble. In other cases, a lack of quality may lead to the firm being unable to deliver on time, thereby loosing money.

Adaptation costs may arise as well, if the firm must rework the goods in order to put them into the production system. This is an indirect cost because no money is actually paid to the supplier, but the firm might have been able to skip this adaptation step if they could find a different supplier.

New machines may require retraining of workers, and this is an indirect cost related to the purchase as well. Equally so are service and maintenance costs, and also purchases of new machines often incur an expense regarding spare parts and storage of these. This expense ties up capital in the firm, and this capital costs money.

Indirect costs linked to the purchasing function

The purchasing department is often a considerable cost in itself. Salaries for the employees, travel expenses, telephone bills, office equipment and other administrative costs should all be counted among the indirect costs of the purchasing function, but it is usually not useful to link these costs to specific transactions. Because these costs are so easy to see, it may be tempting to decentralize the function, but this only serves to camouflage the costs. Instead of linking them to the purchasing department, they will be spread out on the budget of other divisions and/or functions. We should again remind our reader that although the department itself can disappear, the purchasing function is generic and must always be performed by someone.

Table 17.1 Common purchasing costs

Direct costs	*Transaction-specific indirect costs*	*Indirect costs linked to purchasing as a whole*
Price (discounts)	Transportation costs Customs costs Lack of quality Adaptation costs Training costs	Salaries Travel expenses Other administrative costs

17.4.2 Productivity improvements through reduced purchasing costs

We can use Table 17.1 as a checklist when we talk about reductions in purchasing costs. As we have shown before, price is a task that belongs to the operational, external part of the purchasing function, and it is often negotiated by the purchasing department. Reducing prices has been the main way of reducing purchasing costs for a long time, and the pressure toward price reductions is still one of the most prevalent tendencies among purchasers. Reduced prices can be achieved in many ways. Rebates and discounts based on large quantities are common. Another way is to start bidding rounds where the good/service is auctioned to the lowest bidder.

Reduced prices are not without problems, though. Reduced prices may mean a corresponding reduction in quality. Accepting this could incur indirect costs that are much larger than the reduction in direct costs, and purchasers should thus be very careful when bargaining for price cuts or choosing among bidders. Rebates based on quantity are often a better way of reducing prices than accepting lower quality goods for a lower price. However, the price of keeping an increased stock of raw materials must then also be considered.

Cost reductions in indirect costs can also be effective. Philosophies such as Just-In-Time (JIT), Total Quality Management (TQM), strategic alliances and Porterism (Porter, 1980; 1985; 1990) all work on the logic that indirect costs can be reduced through limited cooperation between suppliers and buyers. An example is JIT, which implemented properly will result in savings due to reduced storage requirements, little or no tie-up of capital in raw materials and possibly a reduction of quality problems. Similarly, TQM should reduce costs incurred because of bad quality. Strategic alliances may ensure a steady supply of key raw materials and Porterism can lead to a better control with, and knowledge about, the supplier market, thereby enabling the firm to choose better suppliers.

The problem is that reduction of indirect costs requires a coordinated effort between several functions. To add to the confusion, some of the functions involved are split between different departments as well. As stated before, this is often the case with the purchasing department, which normally has little influence on the strategic side of the purchasing function. Thus, capitalizing on indirect costs would normally mean that the top managers would have to initiate the necessary activities, and to coordinate the work while it is being done. Naturally, such reductions would be easier in firms where the purchasing function has been charged with strategic activities.

17.4.3 Network theory and the purchasing function

So far, we have followed traditional economic theory in our discussion of the purchasing function. This theory focuses on the firm as the interesting unit of analysis. An alternative would be to choose the network as our level of analysis. During the 70's and 80's, theory based on the dyad and the triad and even larger networks of connected firms as the unit of analysis, have surfaced. We usually call this theory 'network theory', and some important works within this tradition are Snehota (1990) and Gadde and Håkansson (1993). For our use it is important to note that these authors stress heavily the importance of cooperation between key actors in the marketplace. Traditional arms-length transactions as in classical and neo-classical theory or strategic alliances as in strategic theory and Porterism, are not strong enough to ensure survival.

Key functions in achieving these close bonds between firms that are so typical in network theory are the marketing/sales functions and the purchasing function. An example in Snehota (1990), claims that a firm's exchange activities are even more important than their productive functions, and that the functions within a firm responsible for these exchange functions, are thus of critical value. Within these functions, network theory emphasizes the importance of the strategic activities.

Utilizing the theoretical basis of network theories enables us to understand better how we can get at the indirect costs mentioned earlier. Closer cooperation between buyer and supplier is seen as critical, and the relation is emphasized over the transaction. Again though, the problem lies in the organization of the purchasing function. Which part of the purchasing function should organize and coordinate such cooperation? The purchasing department does not always have the power to do so being confined to operational activities, and the top managers have a lot of other things on their minds as well.

Network theory can contribute to an understanding on how to reduce indirect costs, but this alone would not be enough to focus on this tradition. Except in sheer explanatory power, there is little that can be done with indirect costs in the network perspective, that cannot be exploited in the more traditional perspectives as well. To really discover the benefits of network theory, we should go even deeper and establish a concept called network benefits that relate the advantages than can be achieved through collaboration and collective action, to a certain way of handling organizational and strategic problems.

Although we do not want to present network theories in detail, we recommend the 'Uppsala school' (Håkansson and Johansson, 1992; Gadde and Håkansson, 1993; Håkansson and Snehota, 1993) as an example of literature in certain areas where interesting forms of network benefits are described in more detail.

The most important of these areas are in research and development. The Uppsala school emphasizes technological development within the dyad as an important bond between supplier and buyer. Suppliers can give considerable contributions to the buying firm when it comes to technological development (Gadde and Håkansson, 1993).

Another important area is adaptation of the production process (Dubois and Håkansson, 1993). One of the three main links between supplier and buyer is the link between production activities, and there is usually considerable benefit to be achieved from improving this linkage. This can be done through adaptations of production equipment and processes, through adaptation of the goods traded between buyer and supplier and through adaptation of administrative procedures. The physical relocation of production facilities could be viewed as an extreme form of adaptation within such a relationship.

Unfortunately, the utilization of network benefits is not without problems either. Network theory presupposes close connections between buyer and seller, and these consume considerable investment. Equally in regard to time, personnel and money, a network linkage needs access to resources. Such close and resource-consuming linkages are not possible with each and every supplier, and neither are they wanted. This means that the buyer must reduce the number of key suppliers to a minimum. Network benefits can then be realized in a cooperation between the buyer and five to fifteen key suppliers. For the rest of the suppliers; more traditional, arms-length relationships will have to suffice.

Another problem is that the close linkage tends to lead to a considerable amount of dependency on the supplier. It is not easy to disentangle oneself from such a link with a supplier, and this dependency makes the buyer more vulnerable to changes in the economic conditions that surround the dyad.

A third problem is that network benefits can only be realized if there exists a considerable amount of trust between the partners, as well as a history of cooperation. Thus, it takes time to establish a network linkage, and it is not something that a company can just decide to establish. Rather, close linkages tend to evolve and grow from more traditional arms-length relationships.

REFERENCES

Ackerman, R.B., Coleman, R.J., Leger, E, and MacDorman, J.C. (1987) *Process Quality Management & Improvement Guidelines*, Publication Center AT&T Laboratories, Indianapolis.

Akao, Y. (editor) (1990) *Quality Function Deployment: Integrating Customer Requirements into Product Design*, Productivity Press, Cambridge, Mass.

Andersen, B. (1994a) Benchmarking in Norwegian industry & relationship benchmarking, *Proceedings from IFIP WG 5.7 Workshop Benchmarking - Theory and practice*, Trondheim, Norway.

Andersen, B. (1994b) A benchmarking process model - the benchmarking wheel. *Proceedings from The 10th International Conference of the Israel Society for Quality*, Jerusalem, Israel.

Andersen, B. and Pettersen, P.G. (1994a) *Håndbok i benchmarking*, TOPP, Trondheim, Norway.

Andersen, B. and Pettersen, P.G. (1994b) The basis of benchmarking: What, how, when, and why. *Proceedings from 1994 Pacific Conference on Manufacturing*, Djakarta, Indonesia.

APQC: American Productivity & Quality Center (1993) *Basics of Benchmarking*, APQC, Houston, Texas.

APQC: American Productivity & Quality Center (1993) *Tools and Techniques for Effective Benchmarking Studies*, APQC, Houston, Texas.

Atwood, C.L. *et al.* (1993) Integration of rapid prototyping into design and manufacturing. *Proceedings from IFIP TC5 Towards World Class Manufacturing*, Phoenix, Arizona.

Aune, A. (1985) *Kvalitetssirkler: Problemløsningsgrupper for personlig vekst, kvalitet og produktivitet*, Universitetsforlaget, Oslo, Norway.

Aune, A. (1990) Slurv koster oss 90 milliarder. *Dagens Næringsliv*, 7.7.1990, Oslo, Norway, pp 24.

Aune, A. (1993) *Kvalitetsstyrte bedrifter*, Ad Notam Gyldendal, Oslo, Norway.

Baily, P. and Farmer, D. (1990) *Purchasing Principles & Management*, 6th ed, Pitman Publishing, London, Great Britain.

Bauer, Bowden, Browne, Duggan and Lyons (1991) *Shop Floor Control Systems: From Design to Implementation*, Chapman & Hall, London, Great Britain.

Bjørke, Ø. and Myklebust, O. (editors) (1992) *IMPPACT Integrated Modelling of Products and Processes using Advanced Computer Technologies. ESPRIT II - CIM III.1.3.1 Project no. 2165*, Tapir, Trondheim, Norway.

Blom, H. and Hovda, J.A. (1985) *Innkjøp: teoristudie og feltundersøkelse om organisasjon, strategi og innkjøpsprosess*, Diploma thesis, Norwegian School of Management, Oslo, Norway.

Bø, K. (1989) *CIM: datamaskinintegrert produksjon*, Universitetsforlaget, Oslo, Norway.

Bondlid, K.O. (1971) *Innkjøp: Den administrative og den utførende side av innkjøpsfunksjonen samt tilgrensende funksjoner*, Norsk Innkjøps- og Materialadministrasjonsforbund (NIMA), Oslo, Norway.

Borthick, A.F. and Roth, H.P. (1993) Accounting for time: re-engineering business processes to improve responsiveness, *Journal of Cost Management*, **7** (3).

Bredrup, H. (1994a) Measurement systems based on standard costing inhibit JIT implementation, *Proceedings from Eighth International Working Seminar on Productions Economics*, Igls/Innsbruck, Austria.

Bredrup, H. (1994b) The ability to achieve real quality through utilization of the ISO 9000 standard, *Proceedings from EOQ'94*, Lisboa, Portugal.

Bredrup, H. (1994c) Standard illusions: ISO 9000 as an alibi for quality, *European Quality*, **1** (5).

Bredrup, H. and Bredrup, R. and Rolstadås, A. (1994) PMS contribution to improved enterprise productivity, *Proceedings from IFIP WG5.7 Working Conference on Evaluation of Production Management Methods*, Porto Alegre/Gramado, Brasil.

Bredrup, H., Bredrup, R. and Estensen, L. (1994) Factors influencing effectiveness of investments in modern manufacturing technology, *Proceedings from Automation '94*, Taipei, Taiwan, R.O.C.

Bredrup, H. and Bredrup, R. (1994a) Benchmarking as performance reference in a performance management model, *Proceedings from IFIP WG 5.7 Workshop Benchmarking - Theory and practice*, Trondheim, Norway.

Bredrup, H. and Bredrup, R. (1994b) Quality function deployment as a tool for strategic planning, *Proceedings from 1994 Pacific Conference on Manufacturing*, Djakarta, Indonesia.

Bredrup, R. (1994) PMS factors influencing on enterprise profitability, *Proceedings from Eighth International Working Seminar on Productions Economics*, Igls/Innsbruck, Austria.

Brenden, M. (1990) *Innkjøp - et strategisk satsningsområde*, Norsk Harvard, 1. kvartal 1990, pp. 82-83.

Brimson, J.A. (1991) *Activity accounting. An Activity-Based Costing Approach*, John Wiley & Sons, Boston, Mass.

BS 3811 (1984) *British Standard*, Great Britain.

Bullinger, H.J., Fremerey, F. and Fuhrberg-Bauman, J. (1993) Innovative Production Structures - Precondition for a Customer-oriented Production Management, *Production Research 1993* (ed. V. Orpana and A. Lukka), Elsevier, Amsterdam, Holland.

Burbidge, J.L. (1979) *Group Technology in the Engineering Industry*, Mechanical Engineering Publications Ltd, London, Great Britain.

Burr, J.T. (1990a) The tools of quality: Part I: going with the flow(chart), *Quality Progress*, June 1990.

Burr, J.T. (1990b) The tools of quality: Part VI: Pareto charts, *Quality Progress*, November 1990.

Burr, J.T. (1990c) The tools of quality: Part VII: scatter diagrams, *Quality Progress*, December 1990.

Byrne, P.M. and Markham, W.J. (1991) *Achieving Customer Satisfaction Breakthroughs: Improving Quality and Productivity in the Logistics Process*, Council of Logistics Management, Chicago, Ill.

Cammish, R. and Keough, M. (1991) A strategic role for purchasing. *The McKinsey Quarterly*, (3), pp 22-39.

Carr, D.K. (1993) Managing for effective business process redesign, *Journal of Cost Management*, **7** (3).

Chase, R.B. and Aquiliano, N.J., (1992) *Production and Operations Management: a Life Cycle Approach*, Irwin, Homewood, Ill.

Christiansen, T. (1993) *Modelling Efficiency and Effectiveness of Coordination in Engineering Design Teams*, DnV, Oslo, Norway.

Clark F.O. and Lorenzoni A.B. (1985) *Applied Cost Engineering*, Marcel Dekker, New York, N.J.

Cooper, R, (1988) The rise of activity-based costing - Part one: What is an activity-based cost system? *Journal of Cost Management*, Summer 1988. pp. 45-54.

Cravens, D.W. and Hoffman, L.M. (1977) Analyzing the supplier: reversing the marketing process. *Management Review*, July 1977, pp. 47-54.

Cummings, T.G. and Huse, E.F. (1989) *Organization Development and Change* (fourth edition), West Publishing Company, St. Paul, Minnesota.

Davenport, T.H. (1993) *Process Innovation: Re-engineering Work Through Information Technology*, Ernst & Young. USA.

Deal, T.E. and Kennedy, A.A., (1982), *Corporate Culture: The Rites and Rituals of Corporate Life*, Addison Wesley, Reading, Mass.

Deming, W.E. (1986) *Out of the Crisis: Quality, Productivity and Competitive Position*, Cambridge University Press, Cambridge, Mass.

Dertouzos, M.L., Lester, R.K., Solow, R.M. and The MIT Commission on Industrial Productivity (1989) *Made in America: Regaining the Productive Edge*, The MIT Press, Cambridge, Massachusetts.

Drucker, P.F. (1992) There's more than one kind of team, *Target*, **8** (5)

Dubois, A. and Håkansson, H. (1993) Relationships as Activity Links. Manuscript for *The 9th IMP Conference*, Bath, Great Britain.

Ericsson, D. and Persson, G. (1986) *Materialadministrasjon. Et företagsledaransvar*, Liber Tryck, Stockholm, Sweden.

Farmer, D. (editor) (1985) *Purchasing Management Handbook*, Gower Publishing, Aldershot, Great Britain.

Feigenbaum, A.V. (1983) *Total Quality Control*, Mcgraw-Hill, Inc. New York, N.J.

Fellers, G. (1992) *The Deming Vision: SPC/TQM For Administrators*, ASQC Quality Press, Milwaukee, Wisconsin.

Gadde, L.E. and Håkansson, H. (1993) *Professional Purchasing*, Routledge, London, Great Britain.

Galbraith, J. R. (1977) *Organization design*, Addison-Wesley, Reading, Mass.

Gilmore, J.H. (1993) Reengineering for mass customization. *Journal of Cost Management*, **7** (3) pp. 22-29.

Govers, C.P.M. (1994) What and how of quality function deployment (QFD), *Proceedings from Eight International Working Seminar on Productions Economics*. Igls/Innsbruck, Austria.

Granli O., Hetland P.W. and Rolstadås A. (1986) *Applied Project Management*, Tapir Publishers, Trondheim, Norway.

Håkansson, H. (1988) *Corporate Technological Behaviour*, Routledge, London, Great Britain.

Håkansson, H. and Johanson, J. (1992) A model of industrial networks, in *Industrial Networks. A New View of Reality*, (eds Axelson, Bjørn and Easton, Geoffrey), Routledge, New York

Håkansson, H. and Snehota, I. (1993) *Business Networks*, Manuscript from a Ph.D. course (Networks) at the Norwegian Institute of Technology, Spring 1993, Trondheim, Norway.

Hammer, M and Champy, J. (1993) *Re-engineering the Corporation: A Manifesto for Business Revolution*, Harper Business, New York, N.J.

Harrington, H.J. (1991) *Business Process Improvement: The Breakthrough Strategy for Total Quality, Productivity, and Competitiveness*, McGraw-Hill, New York, N.J.

Harrington, H.J. (1993) Beyond blind faith, *European Quality*, June 1993, pp. 60-67.

Hayes, R.H. and Wheelwright, S.C. (1984) *Restoring our Competitive Edge: Competing through Manufacturing*, John Wiley & Sons, Boston, Mass.

Howell, R.A. *et al.* (1987) *Management Accounting in the New Manufacturing Environment*, National Association of Accountants, Montvale, N.J.

Imai, M. (1986) *KAIZEN: The Key to Japan's Competitive Success*, Random House Business Division, New York, N.J.

IMS Globeman 21 (1994) *Intelligent Manufacturing Systems. Test Case Globeman 21. Global Manufacturing Towards the Twenty First Century*, IMS, Great Britain.

ISO 9000-9004 (1987) *International Standard Organization*, Great Britain.

ISO 9000-9004 (1989) *International Standard Organization*, Norway.

Istvan, R.L. (1992) A new productivity paradigm for competitive advantage, *Strategic Management Journal*, **13**, pp. 525-537.

Jaikumar, R. (1988) *From Filing and Fitting to Flexible Manufacturing: A Study in the Evolution of Process Control*, Harvard Business School Press, Boston, Mass.

Jauch, R.L. and Glueck, W.F. (1988) *Business Policy and Strategic Management*, McGraw-Hill, New York, N.J.

Jewkes, E. and Power, M. (1993) A microeconomic analysis of investments in Just-in-Time manufacturing, *International Journal of Production Economics*, **29**, pp. 313-321

Johnson, H.T. and Kaplan, R.S. (1987) *Relevance Lost: The Rise and Fall of Management Accounting*, Harvard Business School Press, Boston, Mass.

Juran, J.M. (1979) *Quality Control Handbook* (Third edition), McGraw Hill, New York , N.J.

Juran, J. M. (1988) *Juran on Planning for Quality*, Free Press, New York, NJ

Juran, J.M. (1989) *Juran on Leadership for Quality: An Executive Handbook*, Free Press, New York, N.J.

Juran Institute (1990) The Tools of Quality: Part V: Check Sheets, *Quality Progress*, October 1990.

Kaplan, R.S. (editor) (1990) *Measures for manufacturing excellence*, Harvard Business School Press, Boston, Mass.

Kim, J.S. and Miller, J.G. (1992) Challenges for building the value factory: key findings from the 1992 U.S. Manufacturing Futures Survey, *Operations Management Review*, **9** (3), pp. 1-21.

King, A.M. (1993) Green dollars and blue dollars: the paradox of cost reduction, *Journal of Cost Management*, **7** (3), pp. 44-52.

Kochan, A. (1986) *Implementing CIM*, IFS Publicatons Ltd. Kempston, Great Britain.

Kohler, E. (1952) *Dictionary for Accountants*, Prentice-Hall, Englewood Cliffs, N.J.

Kotler, P. (1988) *Marketing management - analysis, planning, implementation and control*, Prentice-Hall, Englewood Cliffs, N.J.

Kristiansen, T. (1991) *Håndbok i innkjøp*, Norsk innkjøps- og materialadministrasjons-forbund (NIMA), Oslo, Norway.

Kroken, O. and Pedersen, A.C. (1992) *Innkjøpsfunksjonen og innkjøpsstrategier. En teoristudie og case-undersøkelse*. Project Thesis, Norwegian Institute of Technology, Trondheim, Norway.

Kume, H. (1992) *Statistical Methods for Quality Improvement*, Association for Overseas Technical Scholarships (AOTS), Tokyo, Japan.

Lawlor, A. (1985) *Productivity Improvement Manual*, Gower Publishing Co., Aldershot, Great Britain.

Lockyer K. (1984) *Critical Path Analysis*, Pitman Publishing, London, Great Britain.

Lynch, R.L. and Cross, K.C. (1991) *Measure up! Yardstick for Continuous Improvement*, Blackwell Business, Cambridge, MA.

Miller, J. and Roth, A. (1988) *Manufacturing Strategies: Executive Summary of the 1988 North American Futures Survey*, Boston University Manufacturing Roundtable, Boston, Mass.

Mizuno, S. (editor) (1988) *Management for Quality Improvement: The 7 New QC Tools*, Productivity Press, Cambridge, Massachusetts.

Mortensen R. (1991) *Activity-Based Management*, Coopers & Lybrand Konsulenter A/S, Denmark.

Moseng, B. and Bredrup, H. (1993) A methodology for industrial studies of productivity performance. *Journal of Production Planning and Control*, **4** (3).

Myklebust, O. (1992) *Standardisation of Information Exchange in Offshore Projects - STEP in Offshore Standardisation*, Trondheim, Norway.

Nakajima, S. (1989) *TPM Development Program: Implementing Total Productive Maintenance*, Productivity Press, Cambridge, Mass.

Neely, A. *et. al.* (1994) Performance measurement system design - Should process based approaces be adapted? *Proceedings from Eight International Working Seminar on Production Economics*, Igls/Insbruck, Austria.

Newendorp. P.O. (1975) *Decision Analysis for Petroleum Exploration*, PennWell Publishing Company, Tulsa, Okla.

O'Neal C. (1992) Understanding your customer: the most critical to customer delight. *Target*, **8** (5).

O'Neal C. and Bertrand K. (1991) *Developing a Winning JIT Marketing Strategy: The Industrial Marketer's Guide*, Prentice-Hall, Englewood Cliffs, N.J.

Oakland, J. (1991) *Total Quality Management*, Butterworth-Heinemann Ltd, Jordan Hill, Oxford, Great Britain.

Ordione, G. S. (1979*) MBO II: A System of Managerial Leadership for the 80s*, Fearon Pitman Publishers, Belmont, California.

Orlicky, J. (1975) *Materials Requirements Planning: the New Way of Life in Production and Inventory Management*, McGraw-Hill, New York, N.J.

Owen, J. (1993). *STEP: An Introduction*, Information Geometers, Winchester, Great Britain.

Park, C. S. and Sharp-Bette, G. P. (1990) *Advanced Engineering Economics*, John Wiley, New York, N.J.

Pedersen, A.C. and Torvatn, T. (1993) *Temainnlegg om innkjøp*, Manuscript, Norwegian Institute of Technology, Trondheim, Norway.

Peters, T. J. and Waterman Jr, R. H. (1982) *In Search of Excellence: Lessons from America's Best-run Companies*, Harper & Row, New York, N.J.

Porter, M.E. (1980) *Competitive Strategy: Techniques for Analyzing Industries and Competitors*, Free Press, New York, N.J.

Porter, M.E. (1985) *Competitive Advantage: Creating and Sustaining Superior Performance*, Free Press, New York, N.J.

Porter, M.E. (1990) *Competitive Advantage of Nations*. The MacMillan Press Ltd, London, Great Britain.

Prytz, K. (1993) Europa og USA føler Japan på pulsen. *Teknisk Ukeblad* (38) pp.34-35.

Prytz, K. (1994) Business Modeling for Information Flow Control in Manufacturing. *Proceedings from Automation '94*, Taipei, Taiwan, R.O.C.

Rembold, U. *et al.* (1993) *Computer Integrated Manufacturing and Engineering*. Addison-Wesley, Wokingham. Great Britain.

Remer, D.S., Stokdyk, S.B. and Driel, M. van (1993) Survey of project evaluation techniques currently used in industry. *International Journal of Production Economics*, **32**, pp. 103-115.

Rolstadås, A. (1986) *Production Management Systems, STF17A86012.* Trondheim, Norway.

Rolstadås, A. (1988) *Cost Effectiveness and Consciousness in Exploration, Development and Production*, SINTEF, Trondheim, Norway.

Rolstadås, A. (1990) *Praktisk prosjektstyring*, Tapir Publishers, Trondheim, Norway.

Rolstadås, A. (1991) *A Conceptual Model for One-of-a-Kind Production*, Institutt for Verkstedteknikk, NTH, Trondheim, Norway.

Sakurai, M. (1990) The influence of factory automation on management accounting practices: a study of Japanese companies, in *Measures for Manufacturing Excellence*, (ed Robert S. Kaplan), Harvard Business School Press, Boston, Mass.

Scheuing, E.E. (1989) *Purchasing Management*, Prentice-Hall International Editions, Englewood Cliffs, N.J.

Sink, D.S. and Tuttle, T.C. (1989) *Planning and Measurement in your Organization of the Future*, Industrial Engineering and Management Press, Norcross.

Sink. D.S. (1985) *Productivity Management: Planning, Measurement and Evaluation, Control and Improvement*, John Wiley & Sons, New York, N.J.

Snehota, I. (1990) *Notes on a Theory of a Business Enterprise,* Doctoral dissertation. Uppsala University, Uppsala, Sweden.

Solberg, J.J. (1989) Production Planning and Scheduling in CIM, *Information Processing 1989*, (ed. G.X. Ritter), Elsevier Science Publisher, pp 919-925, Holland.

Spekman, R.E. (1988) Strategic supplier selection: understanding long-term buyer relationships. *Business Horizons*, July-August 1988, pp 75-81.

Stalk, G. Jr. and Hout, T.M. (1990) *Competing Against Time: How Time-Based Competition is Reshaping Global Markets*, The Free Press, New York, N.J.

Timmermans, P. (1993) *Modular design of information systems for shop floor control*, Eindhoven, The Netherlands.

Watson, Gregory H. (1992) *The Benchmarking Workbook: Adapting Best Practices for Performance Improvement*, Productivity Press, Cambridge, Mass.

Webster's Desk Dictionary of the English Language (1990) Gramercy Books, New York, N.J.

Wiendahl, H.P. (1987) *Belastungsorientierte Fertigungssteuerung*, Carl Hanser Verlag, Munchen, Germany.

Wille, G. (1992) *Fra bedrift til bedrift*, Universitetsforlaget, Oslo, Norway.

Williamson, O.E. (1985) *The Economics of Capitalism*, The Free Press, New York, N.J.

Womack J.P., Jones D.T. and Roos D. (1990) *The Machine that Changed the World*, Macmillan Publishing Company, New York, N.J.

INDEX

GPSR Compliance

The European Union's (EU) General Product Safety Regulation (GPSR) is a set of rules that requires consumer products to be safe and our obligations to ensure this.

If you have any concerns about our products, you can contact us on ProductSafety@springernature.com

In case Publisher is established outside the EU, the EU authorized representative is:

Springer Nature Customer Service Center GmbH
Europaplatz 3
69115 Heidelberg, Germany

Batch number: 10091745

Printed by Printforce, the Netherlands